Yamaha XS250, 360 & 400 Twins Owners Workshop Manual

by Mansur Darlington

with an additional Chapter on the later models

by Pete Shoemark

Models covered

XS250 P. 248cc. UK 1977 to 1979
XS250. 248cc. UK 1978 to 1981
XS250 SE. 248cc. UK 1980 to 1984
XS250 C. 248cc. UK 1981 to 1982
XS360 C. 358cc. US 1975 to 1976
XS360 2D. 358cc. US 1976 to 1977
XS400. 391cc. UK 1977 to 1980
XS400 SE. 391cc. UK 1980 to 1983
XS400 D. 391cc. US 1976 to 1977

XS400 E. 391cc. US 1977 to 1978
XS400 2E. 391cc. US 1977 to 1978
XS400 F. 391cc. US 1978 to 1979
XS400 2F. 391cc. US 1978 to 1979
XS400 G Special II. 391cc. US 1979 to 1980
XS400 SG Special. 391cc. US 1979 to 1980
XS400 H Special II. 391cc. US 1980 to 1981
XS400 SH Heritage. 391cc. US 1980 to 1981
XS400 SJ Heritage. 391cc. US 1981 to 1982

ISBN 978 1 85010 248 9

Printed in the UK   (378-10S8)

ABCDE
FGHIJ
KLMNO
PQR
3

British Library Cataloguing in Publication Data

A catalogue record for this book is available from the British Library

Library of Congress Control Number 86-81748

Haynes Publishing
Sparkford Nr Yeovil
Somerset BA22 7JJ England

Haynes Publications Inc
859 Lawrence Drive
Newbury Park
California 91320 USA

Acknowledgements

Our grateful thanks are due to Fran Ridewood and Co. of Wells, Somerset, who supplied the machine featured in this manual, and to Mitsui Machinery sales (UK) Limited, who gave permission to use the line drawings reproduced throughout this manual and who supplied the colour transparency of the machine featured on the front cover.

Our thanks are also due to Jim Patch of Yeovil Motorcycle Services who supplied us with invaluable technical information when preparing the manual.

Martin Penny gave considerable assistance with the strip-down and rebuilding necessary for the photographic sequences. Les Brazier arranged and took the photographs and Jeff Clew edited the text.

We would also like to thank the following Companies who kindly supplied advice and information relating to their products; The Avon Rubber Company; NGK Spark Plugs (UK) Ltd; and Renold Limited.

About this manual

The author of this manual has the conviction that the only way in which a meaningful and easy to follow text can be written is first to do the work himself, under conditions similar to those found in the average household. As a result, the hands seen in the photographs are those of the author. Unless specially mentioned, and therefore considered essential, Yamaha special service tools have not been used. There is invariably some alternative means of loosening or removing a vital component when service tools are not available, but risk of damage should always be avoided.

Each of the six Chapters is divided into numbered sections. Within these section are numbered paragraphs. Cross reference throughout the manual is quite straightforward and logical. When reference is made 'See Section 6.10' it means Section 6, paragraph 10 in the same Chapter. If another Chapter were meant, the reference would read, for example, 'See Chapter 2, Section 6.10'. All the photographs are captioned with a section/paragraph number to which they refer and are relevant to the Chapter text adjacent.

Figures (usually line illustrations) appear in a logical but numerical order, within a given Chapter. Fig. 1.1 therefore refers to the first figure in Chapter 1.

Left-hand and right-hand descriptions of the machines and their components refer to the left and right of a given machine when the rider is seated normally.

Motorcycle manufacturers continually make changes to specifications and recommendations, and these, when notified, are incorporated into our manuals at the earliest opportunity.

We take great pride in the accuracy of information given in this manual, but motorcycle manufacturers make alterations and design changes during the production run of a particular motorcycle of which they do not inform us. No liability can be accepted by the authors or publishers for loss, damage or injury caused by any errors in, or omissions from, the information given.

Contents

Right-hand view of 1977 Yamaha XS400

Introduction to the Yamaha XS 250, 360 and 400 twins

Although the history of Yamaha can be traced back to the year 1887, when a then very small company commenced manufacture of reed organs, it was not until 1954 that the company became interested in motor cycles. As can be imagined, the problems of marketing a motor cycle against a background of musical instrument manufacture were considerable. Some local racing successes and the use of hitherto unknown bright colour schemes helped achieve the desired results and in July 1955 the Yamaha Motor Company was established as a separate entity, employing a work force of less than 100 and turning out some 300 machines a month.

Competition successes continued and with the advent of tasteful styling that followed Italian trends, Yamaha became established as one of the world's leading motor cycle manufacturers. Part of this success story is the impressive list of Yamaha 'firsts' – a whole string of innovations that include electric starting, pressed steel frame, torque induction and 6 and 8 port engines. There is also the 'Autolube' system of lubrication, in which the engine-driven oil pump is linked to the twist grip throttle, so that lubrication requirements are always in step with engine demands.

With the changing world opinion on pollution and fuel economy, Yamaha, who built their worldwide reputation on the manufacture of two-stroke machines, have moved into the four-stroke market, to compete with the other large motorcycle manufacturers who have done likewise.

The XS range of four-strokes, of which the XS250, 360 and 400 are the latest additions, can be considered as the first generation of four-strokes of this marque.

The XS250, 360 and 400 are all of similar construction, having an almost identical engine and sharing the same frame. The 360 cc machine has been available in three different versions comprising the XS360, XS360C and the XS360-2D. The latter model was introduced with economy in mind as it is fitted with drum brakes at both front and rear and is not provided with a starter motor or self-cancelling indicators.

Dimensions and weights

	XS250	XS360	XS400
Overall length	2020 mm (79.52 in)	2045 mm (80.5 in)	2025 mm (79.7 in)
Overall width	730 mm (28.74 in)	845 mm (33.3 in)	845 mm (33.3 in)
Overall height	1070 mm (42.12 in)	1340 mm (52.8 in)	1335 mm (52.6 in)
Wheel base	1330 mm (52.36 in)	800 mm (31.5 in)	815 mm (32.1 in)
Weight	166 kg (365 lbs)	159 kg (350 lbs)	164 kg (361 lbs)

Ordering spare parts

When ordering spare parts for the Yamaha four-stroke twins it is advisable to deal direct with an official Yamaha agent, who will be able to supply many of the items required ex-stock. Although parts can be ordered from Yamaha direct, it is preferable to route the order via a local agent even if the parts are not available from stock. He is in a better position to specify exactly the parts required and to identify the relevant spare part numbers so that there is less chance of the wrong part being supplied by the manufacturer due to a vague or incomplete description.

When ordering spares, always quote the frame and engine numbers in full, together with any prefixes or suffixes in the form of letters. The frame number is found stamped on the right-hand side of the steering head, in line with the forks. The engine number is stamped on the right-hand side of the upper crankcase, immediately below the left-hand carburettor.

Use only parts of genuine Yamaha manufacture. A few pattern parts are available, sometimes at cheaper prices, but there is no guarantee that they will give such good service as the originals they replace. Retain any worn or broken parts until the replacements have been obtained; they are sometimes needed as a pattern to help identify the correct replacement when design changes have been made during a production run.

Some of the more expendable parts such as spark plugs, bulbs, tyres, oils and greases etc., can be obtained from accessory shops and motor factors, who have convenient opening hours, and can often be found not far from home. It is also possible to obtain parts on Mail Order basis from a number of specialists who advertise regularly in the motor cycle magazines.

Frame number location

Engine number location

Safety first!

Professional motor mechanics are trained in safe working procedures. However enthusiastic you may be about getting on with the job in hand, do take the time to ensure that your safety is not put at risk. A moment's lack of attention can result in an accident, as can failure to observe certain elementary precautions.

There will always be new ways of having accidents, and the following points do not pretend to be a comprehensive list of all dangers; they are intended rather to make you aware of the risks and to encourage a safety-conscious approach to all work you carry out on your vehicle.

Essential DOs and DON'Ts

DON'T start the engine without first ascertaining that the transmission is in neutral.

DON'T suddenly remove the filler cap from a hot cooling system – cover it with a cloth and release the pressure gradually first, or you may get scalded by escaping coolant.

DON'T attempt to drain oil until you are sure it has cooled sufficiently to avoid scalding you.

DON'T grasp any part of the engine, exhaust or silencer without first ascertaining that it is sufficiently cool to avoid burning you.

DON'T allow brake fluid or antifreeze to contact the machine's paintwork or plastic components.

DON'T syphon toxic liquids such as fuel, brake fluid or antifreeze by mouth, or allow them to remain on your skin.

DON'T inhale dust – it may be injurious to health (see *Asbestos* heading).

DON'T allow any spilt oil or grease to remain on the floor – wipe it up straight away, before someone slips on it.

DON'T use ill-fitting spanners or other tools which may slip and cause injury.

DON'T attempt to lift a heavy component which may be beyond your capability – get assistance.

DON'T rush to finish a job, or take unverified short cuts.

DON'T allow children or animals in or around an unattended vehicle.

DON'T inflate a tyre to a pressure above the recommended maximum. Apart from overstressing the carcase and wheel rim, in extreme cases the tyre may blow off forcibly.

DO ensure that the machine is supported securely at all times. This is especially important when the machine is blocked up to aid wheel or fork removal.

DO take care when attempting to slacken a stubborn nut or bolt. It is generally better to pull on a spanner, rather than push, so that if slippage occurs you fall away from the machine rather than on to it.

DO wear eye protection when using power tools such as drill, sander, bench grinder etc.

DO use a barrier cream on your hands prior to undertaking dirty jobs – it will protect your skin from infection as well as making the dirt easier to remove afterwards; but make sure your hands aren't left slippery. Note that long-term contact with used engine oil can be a health hazard.

DO keep loose clothing (cuffs, tie etc) and long hair well out of the way of moving mechanical parts.

DO remove rings, wristwatch etc, before working on the vehicle – especially the electrical system.

DO keep your work area tidy – it is only too easy to fall over articles left lying around.

DO exercise caution when compressing springs for removal or installation. Ensure that the tension is applied and released in a controlled manner, using suitable tools which preclude the possibility of the spring escaping violently.

DO ensure that any lifting tackle used has a safe working load rating adequate for the job.

DO get someone to check periodically that all is well, when working alone on the vehicle.

DO carry out work in a logical sequence and check that everything is correctly assembled and tightened afterwards.

DO remember that your vehicle's safety affects that of yourself and others. If in doubt on any point, get specialist advice.

IF, in spite of following these precautions, you are unfortunate enough to injure yourself, seek medical attention as soon as possible.

Asbestos

Certain friction, insulating, sealing, and other products – such as brake linings, clutch linings, gaskets, etc – contain asbestos. *Extreme care must be taken to avoid inhalation of dust from such products since it is hazardous to health.* If in doubt, assume that they *do* contain asbestos.

Fire

Remember at all times that petrol (gasoline) is highly flammable. Never smoke, or have any kind of naked flame around, when working on the vehicle. But the risk does not end there – a spark caused by an electrical short-circuit, by two metal surfaces contacting each other, by careless use of tools, or even by static electricity built up in your body under certain conditions, can ignite petrol vapour, which in a confined space is highly explosive.

Always disconnect the battery earth (ground) terminal before working on any part of the fuel or electrical system, and never risk spilling fuel on to a hot engine or exhaust.

It is recommended that a fire extinguisher of a type suitable for fuel and electrical fires is kept handy in the garage or workplace at all times. Never try to extinguish a fuel or electrical fire with water.

Note: *Any reference to a 'torch' appearing in this manual should always be taken to mean a hand-held battery-operated electric lamp or flashlight. It does **not** mean a welding/gas torch or blowlamp.*

Fumes

Certain fumes are highly toxic and can quickly cause unconsciousness and even death if inhaled to any extent. Petrol (gasoline) vapour comes into this category, as do the vapours from certain solvents such as trichloroethylene. Any draining or pouring of such volatile fluids should be done in a well ventilated area.

When using cleaning fluids and solvents, read the instructions carefully. Never use materials from unmarked containers – they may give off poisonous vapours.

Never run the engine of a motor vehicle in an enclosed space such as a garage. Exhaust fumes contain carbon monoxide which is extremely poisonous; if you need to run the engine, always do so in the open air or at least have the rear of the vehicle outside the workplace.

The battery

Never cause a spark, or allow a naked light, near the vehicle's battery. It will normally be giving off a certain amount of hydrogen gas, which is highly explosive.

Always disconnect the battery earth (ground) terminal before working on the fuel or electrical systems.

If possible, loosen the filler plugs or cover when charging the battery from an external source. Do not charge at an excessive rate or the battery may burst.

Take care when topping up and when carrying the battery. The acid electrolyte, even when diluted, is very corrosive and should not be allowed to contact the eyes or skin.

If you ever need to prepare electrolyte yourself, always add the acid slowly to the water, and never the other way round. Protect against splashes by wearing rubber gloves and goggles.

Mains electricity and electrical equipment

When using an electric power tool, inspection light etc, always ensure that the appliance is correctly connected to its plug and that, where necessary, it is properly earthed (grounded). Do not use such appliances in damp conditions and, again, beware of creating a spark or applying excessive heat in the vicinity of fuel or fuel vapour. Also ensure that the appliances meet the relevant national safety standards.

Ignition HT voltage

A severe electric shock can result from touching certain parts of the ignition system, such as the HT leads, when the engine is running or being cranked, particularly if components are damp or the insulation is defective. Where an electronic ignition system is fitted, the HT voltage is much higher and could prove fatal.

Routine maintenance

Periodic routine maintenance is a continuous process that commences immediately the machine is used and continues until the machine is no longer fit for service. It must be carried out at specified mileage recordings or on a calendar basis if the machine is not used regularly, whichever is the soonest. Maintenance should be regarded as an insurance policy, to help keep the machine in the peak of condition and to ensure long, trouble-free service. It has the additional benefit of giving early warning of any faults that may develop and will act as a safety check, to the obvious advantage of both rider and machine alike.

The various maintenance tasks are described under their respective mileage and calendar headings. Accompanying photos or diagrams are provided, where necessary. It should be remembered that the interval between the various maintenance tasks serves only as a guide. As the machine gets older, is driven hard, or is used under particularly adverse conditions, it is advisable to reduce the period between each check.

For ease of reference each service operation is described in detail under the relevant heading. However, if further general information is required it can be found within the manual in the relevant Chapter.

Although no special tools are required for routine maintenance, a good selection of general workshop tools are essential. Included in the tools must be a range of metric ring or combination spanners, a selection of cross-head screwdrivers, and two pairs of circlip pliers, one external opening and the other internal opening. Additionally, owing to the extreme tight-ness of most casing screws on Japanese machines, an impact screwdriver together with a choice of large or small cross-head screw bits, is absolutely indispensable. This is particularly so if the engine has not been dismantled since leaving the factory.

Weekly or every 200 miles (300 km)

1 Tyre pressures

Check the tyre pressures with a pressure gauge that is known to be accurate. Always check the pressures when the tyres are cold. If the tyres are checked after the machine has travelled a number of miles, the tyres will have become hot and consequently the pressure will have increased, possibly as much as 8 psi. A false reading will therefore always result.

Tyre pressures:	Solo	Pillion or continuous high speed
Front tyre	26 psi (1.8 kg/cm²)	28 psi (2.0 kg/cm²)
Rear tyre	28 psi (2.0 kg/cm²)	33 psi (2.3 kg/cm²)

2 Engine/transmission oil level check

Check the engine/transmission oil level by means of the dipstick integral with the filler cap in the primary drive casing. To ascertain the level, unscrew the cap, wipe the dipstick clean, and then insert the dipstick so that the cap rests on the top edge

Check engine oil level by means of dipstick

Replenish with SAE 20W/50 engine oil

of the casing. Do not screw the cap in to determine the level. If necessary, replenish with SAE 20W/50 engine oil.

3 Hydraulic fluid level (disc brake models only)

Check the level of the hydraulic fluid in the front brake master cylinder reservoir, on the handlebars, and also the rear brake reservoir, behind the right-hand frame side cover. The level in both reservoirs should lie between the upper and lower level marks. During normal service, it is unlikely that the hydraulic fluid level will fall dramatically, unless a leak has developed in the system. If this occurs, the fault should be remedied AT ONCE. The level will fall slowly as the brake linings wear and the fluid deficiency should be corrected, when required. Always use an hydraulic fluid of DOT 3 or SAE J1703 specification, and if possible do not mix different types of fluid, even if the specifications appear the same. This will preclude the possibility of two incompatible fluids being mixed and the resultant chemical reactions damaging the seals.

If the level in either reservoir has been allowed to fall below the specified limit, and air has entered the system, the brake in question must be bled, as described in Chapter 5, Section 17.

4 Control cable lubrication

Apply a few drops of motor oil to the exposed inner portion of each control cable. This will prevent drying-up of the cables between the more thorough lubrication that should be carried out during the 2000 mile/3 monthly service.

5 Rear chain lubrication and adjustment

In order that the life of the rear chain be extended as much as possible, regular lubrication and adjustment is essential.

Intermediate lubrication should take place at the weekly or 200 mile service interval with the chain in situ. Application of one of the proprietary chain greases contained in an aerosol can is ideal. Ordinary engine oil can be used, though owing to the speed with which it is flung off the rotating chain, its effectiveness is limited.

Adjust the chain after lubrication, so that there is approximately 20 mm ($\frac{3}{4}$ in) slack in the middle of the lower run. Always check with the chain at the tightest point as a chain rarely wears evenly during service.

Adjustment is accomplished after placing the machine on the centre stand and slackening the wheel nut, so that the wheel can be drawn backwards by means of the drawbolt adjusters in the fork ends. On drum brake models the torque arm nut must also be slackened during this operation. Adjust the drawbolts an equal amount to preserve wheel alignment. The fork ends are clearly marked with a series of parallel lines above the adjusters, to provide a simple visual check.

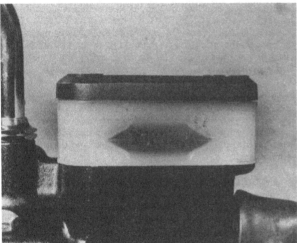

Check fluid level of front and ...

... rear brake fluid reservoirs

Remove cap and diaphragm to replenish

Use aerosol chain lubricant to lubricate chain, on machine

6 Safety check

Give the machine a close visual inspection, checking for loose nuts and fittings, frayed control cables etc. Check the tyres for damage, especially splitting on the sidewalls. Remove any stones or other objects caught between the treads. This is particularly important on the front tyre, where rapid deflation due to penetration of the inner tube will almost certainly cause total loss of control.

7 Legal check

Ensure that the lights, horn and trafficators function correctly, also the speedometer.

Monthly or every 1000 miles (1600 km)

Carry out the checks listed under the weekly/200 mile heading and then complete the following:

1 Final drive chain lubrication

The final drive chain should be removed from the machine for thorough cleaning and lubrication if long service life is to be expected. This is in addition to the intermediate lubrication carried out with the chain on the machine, as described under the weekly/200 mile service heading.

Separate the chain by removing the master link and run it off the sprockets. If an old chain is available, interconnect the old and new chain, before the new chain is run off the sprockets. In this way the old chain can be pulled into place on the sprockets and then used to pull the regreased chain into place with ease.

Clean the chain thoroughly in a paraffin bath and then finally with petrol. The petrol will wash the paraffin out of the links and rollers which will then dry more quickly.

Allow the chain to dry and then immerse it in a molten lubricant such as Linklyfe or Chainguard. These lubricants must be used hot and will achieve better penetration of the links and rollers, and are less likely to be thrown off by centrifugal force when the chain is in motion.

Refit the newly greased chain onto the sprocket, replacing the master link. This is accomplished most easily when the free ends of the chain are pushed into mesh on the rear wheel sprocket. The spring link must be fitted so that the closed end faces the normal direction of chain travel.

2 Battery electrolyte level

A conventional lead-acid battery is fitted to all models.

Electrolyte levels are clearly marked on battery

The transparent plastic case of the battery permits the upper and lower levels of the electrolyte to be observed when the battery is lifted from its housing below the dual seat. Maintenance is normally limited to keeping the electrolyte level between the prescribed upper and lower limits and by making sure that the vent pipe is not blocked. The lead plates and their separators can be seen through the transparent case, a further guide to the general condition of the battery.

Unless acid is spilt, as may occur if the machine falls over, the electrolyte should always be topped up with distilled water, to restore the correct level. If acid is spilt on any part of the machine, it should be neutralised with an alkali such as washing soda and washed away with plenty of water, otherwise serious corrosion will occur. Top up with sulphuric acid of the correct specific gravity (1.260 – 1.280) only when spillage has occurred. Check that the vent pipe is well clear of the frame tubes or any of the other cycle parts, for obvious reasons.

3 Adjusting the front and rear brakes (drum brake models only)

The front brake should be adjusted so that there is 5.8 mm (0.2 – 0.3 in) movement, measured between the handlebar lever and the lever stock, before the brake starts to bite. Adjustment may be made either at the cable adjuster on the handlebar lever or the cable adjuster on the brake back plate.

If uneven brake shoe wear reduces the efficiency of the brake, the brake arm link rod should be adjusted as described in Chapter 5, Section 10.

Adjust the rear brake by means of the adjuster nut on the brake rod so that there is approximately 30 mm (1.2 in) vertical movement at the brake pedal foot pad before the brake commences operation.

Three monthly or every 2000 miles (3200 km)

Complete the tasks enumerated in the weekly/200 mile and monthly/1000 mile maintenance schedules, and then carry out the following:

1 Changing the engine/transmission oil

Place a container of more than 3.0 litres (6.4/5.3 US/Imp pints) below the crankcase. Remove the filler cap from the primary drive case and unscrew the oil drain plug from the base of the crankcase. A second plug fitted to the left-hand wall of the crankcase below the alternator cover may also be removed. Drain the oil after the engine has been allowed to reach normal working temperature, preferably after a run; the oil will be thinner when hot and so drain more rapidly and completely.

Replace the drain plugs and refill the engine with 2.0 litres (4.2/3.5 US/Imp pints) of SAE 20W/50 engine oil. Allow the oil to settle for a few moments and then check the level by means of the dipstick in the filler cap. Add more oil, if required.

2 Cleaning the air filter elements

Two identical air filter boxes are fitted, each of which contains a woven fabric air filter element. One box is fitted behind each frame side cover which must be detached to gain access. Remove and clean each filter separately, in the same manner, as follows:

Slacken the air hose/air filter box screw clip and remove the single bolt securing the lower end of the air filter box retaining strap. Hinge the strap up and pull the air filter box from place. The air filter box is a two-piece moulding held together by two screws. After removal of the screws, separate the two box halves and withdraw the element.

The air filter is best cleaned using a compressed air line directed from the inside. If this is not available, the dust should be loosened, using a soft brush, and then blown out, using a tyre pump.

A badly soiled or oily element will restrict the flow of air to the engine and so reduce performance. In addition, the mixture will be richened, increasing the fuel consumption proportionately.

Do not on any account run the machine with the air filter removed or with the air cleaner hoses disconnected. If this precaution is not observed, the engine will run with a permanently weak mixture, which will cause overheating and possible seizure.

3 Control cable lubrication

Lubricate the control cables thoroughly with motor oil or an all-purpose oil. A good method of lubricating the cables is shown in the accompanying illustration, using a plasticine funnel. This method has the disadvantage that the cables usually need removing from the machine. An hydraulic cable oiler which pressurises the lubricant overcomes this problem. Do not lubricate nylon lined cables (which may have been fitted as replacements), as the oil may cause the nylon to swell, thereby causing total cable seizure.

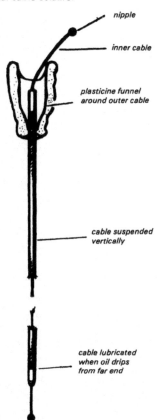

nipple

inner cable

plasticine funnel around outer cable

cable suspended vertically

cable lubricated when oil drips from far end

Oiling control cables

4 General lubrication

Apply grease or oil to the handlebar lever pivots and to the centre stand and prop stand pivots.

5 Wheel condition (spoke type)

Check the spoke tension by gently tapping each one with a metal object. A loose spoke is identifiable by the low pitch noise generated. If any one spoke needs considerable tightening, it will be necessary to remove the tyre and inner tube in order to file down the protruding spoke end. This will prevent the spoke from chafing through the rim band and piercing the inner tube. Rotate the wheel and test for rim runout. Excessive runout will cause handling problems and should be corrected by tightening or loosening the relevant spokes. Care must be taken, since altering the tension in the wrong spokes may create more problems.

6 Cleaning and checking spark plugs

Remove the spark plugs and clean them, using a wire brush. Clean the electrodes using fine emery paper or cloth and then reset the gaps to 0.7 – 0.8 mm (0.028 – 0.030 in) with the correct feeler gauge. Before replacing the plugs, smear the threads with a small amount of graphite grease to aid future removal.

7 Cleaning and adjusting the contact breaker points

To gain access to the contact breaker assembly, it is necessary to remove the cover plate which is held by two cross head screws to the cylinder head. Note that the cover has a paper gasket to prevent the ingress of water.

Remove the alternator inspection cover and rotate the engine until the left-hand points are fully open. Removal of the spark plugs will aid rotation.

Inspect the faces of each of the two sets of contact breaker points. Slight burning or pitting can be removed while the contact breaker unit is in situ on the machine, using a very fine swiss file or emery paper (No. 400) backed by a thin strip of tin. If pitting or burning is excessive the contact breaker unit should be removed for points dressing or renewal. (See Chapter 3, Section 4). After the points have been cleaned, they should be adjusted to restore the correct gap as follows:

Adjustment is effected by slackening the two screws through the plate of the fixed contact breaker point and moving the point either closer to, or further away, from the moving contact until the gap is correct as measured by a feeler gauge. The correct gap with the points FULLY OPEN is 0.3 – 0.4 mm (0.012 – 0.016 in).

Two small projections on the contact breaker base plate permit the insertion of a screwdriver to lever the adjustable point into its correct location. Repeat this operation if there is any doubt about the accuracy of the measurement. Although this adjustment is relatively easy, it is of prime importance.

Carry out the procedure on the right-hand set of contact breakers after turning the engine again so that the points are fully open.

After making contact breaker adjustments, the ignition timing should be checked, and if necessary adjusted, as described under the following heading.

8 Ignition timing

It is recommended that the ignition timing check be carried out statically (with the engine stopped), rather than dynamically using a stroboscope. Although use of a stroboscope will enable checking of the full advance operation, it should be noted that in running the engine, oil will issue from the open crankcase left-hand half with the resultant loss of engine lubrication. To carry out the static timing check proceed as follows.

Start by inspection and setting the contact breakers as described previously. Either drain the engine oil, or prepare to catch oil spillage when the circular inspection cover on the left-hand engine side (4 screws) is removed.

With the cover removed, turn the engine slowly and look through the inspection slot in the casing upper section to identify the timing marks on the alternator rotor periphery. For each cylinder there will be a full advance mark (two single lines close together), followed by the timing mark (LF for left cylinder, RF for right), and then the TDC mark (LT for left cylinder, RT for right). For the static timing check, the LF (left cylinder) and RF (right cylinder) marks are used. Note that the actual mark for alignment purposes is the full scribed line next to the LF or RF (see photo 5.1a in Chapter 3).

Check the left-hand cylinder first. In order to determine the exact moment of contact breaker point separation connect up an audible points checker, multimeter set to the resistance range or a battery and bulb test circuit across the left-hand contact breaker set. This is best done by connecting the equipment between the orange wire from the points and a good earth on the engine. When the points separate the buzzer will sound, meter needle swing or bulb light (as applicable).

Rotate the engine slowly anticlockwise until the left cylinder is just approaching the LF mark on its compression stroke. If the timing is set correctly, the points should separate when the LF mark aligns with the static casing mark at the top of the inspection aperture.

If the timing is incorrect, align the rotor LF mark with the static casing mark, then slacken the two screws which clamp the contact breaker baseplate to the cylinder head. Rotate the complete contact breaker assembly until the equipment indicates that the points have just separated and tighten the baseplate screws. Recheck the timing by rotating the engine backwards about 45° and then forwards again until the LF mark aligns precisely with the index mark. The engine must be rotated one way and then the other to take up any backlash in the timing chain (camshaft chain).

Repeat the timing check procedure on the right-hand contact-breaker, using the RF mark on the alternator rotor and the grey wire from the right-hand contact breaker set. If the timing is incorrect **DO NOT** slacken the two screws previously used for timing. The right-hand contact breaker is mounted on a separate smaller base plate, retained by two screws passing through elongated holes in the plate. The elongated holes allow a limited amount of plate movement for ignition timing.

Disconnect the test circuit from the points and refit all disturbed components. Replenish any lost engine oil.

Before replacing the contact breaker cover apply a small amount of thin oil to the cam lubricating wicks. One or two drops will suffice. Too much oil will contaminate the points faces, causing ignition failure.

9   Adjusting the tick-over speed

The engine tick-over speed may be adjusted by means of the throttle pulley stop screw, which is located between the two carburettors. The screw can be identified by the serrated plastic knob. The correct tick-over speed is 1,200 rpm.

If the two carburettors are not synchronised accurately, refer to Chapter 2 Section 7 for the correct adjustment procedure.

Six monthly or every 4000 miles (6400 km)

Carry out the tasks described in the weekly, monthly and three monthly sections and then carry out the following:

1   Checking and adjusting the valve clearances

The engine must be completely cold before the valve clearances can be checked accurately. Remove the fuel tank, both spark plugs and the alternator inspecton cover (be prepared for some oil loss), also the four inspection caps from the camshaft cover. Rotate the crankshaft anti-clockwise until the left-hand cylinder inlet valve has opened and closed (sunk down and risen again), then rotate the crankshaft further until the alternator rotor LT mark aligns with the crankcase pointer. The left-hand cylinder

Check points gap when in fully open position

Slacken two screws to adjust each set of points

A = LH cylinder timing adjustment screws; B = RH cylinder screws

Lubricate cam wick sparingly

is then at TDC on the compression stroke (both valves closed), with free play at both rockers. The correct clearances are:

Inlet   0.08 – 0.12 mm (0.003 – 0.005 in)
Exhaust 0.16 – 0.20 mm (0.006 – 0.008 in)

Measure the gap by inserting a feeler gauge between the rocker arm adjuster screw and the tip of the valve stem; if the gap is correct a gauge of the required thickness will be a tight slip fit. To adjust the clearance, slacken the adjuster locknut and screw the adjuster in or out as necessary, then hold the screw steady while the locknut is tightened. Tighten the locknut securely but be careful not to overtighten it as this merely distorts the adjuster threads and makes future adjustment very difficult. Recheck the clearance to ensure that the setting has not altered. Repeat for the remaining valve.

When the left-hand cylinder valve clearances are correct, turn the crankshaft forwards until the right-hand cylinder is positioned, as described above, at TDC on the compression stroke (alternator rotor RT mark aligned and free play at both rockers) then check the clearances of the right-hand valves.

Refit the inspection caps and other disturbed components. Top up the engine oil to replace any lost when the alternator inspection cover was removed.

2 Changing the oil filter

During the 4000 mile service, ie at every second oil change, the old oil filter should be removed and discarded, and a replacement element fitted.

The oil filter is contained within a semi-isolated chamber at the front of the crankcase. Access to the element is made by unscrewing the filter cover centre bolt which will bring with it the cover and also the element. Before removing the cover, place a receptacle beneath the engine to catch the engine oil contained in the filter chamber.

When renewing the filter element it is wise to renew the filter cover 'O'ring at the same time. This will obviate the possibility of any oil leaks. Do not overtighten the centre bolt on replacement; the correct torque setting is 1.3 – 1.7 kg m (9.5 – 12.0 lb ft).

The filter by-pass valve, comprising a plunger and spring, is situated in the bore of the filter cover centre bolt. It is recommended that the by-pass valve be checked for free movement during every filter change. The spring and plunger are retained by a pin across the centre bolt. Knocking the pin out will allow the spring and plunger to be removed for cleaning.

Never run the engine without the filter element or increase the period between the recommended oil changes or oil filter changes.

Yearly or every 8000 miles (12 875 km)

Again complete the checks listed under the previous routine maintenance interval headings. The following additional tasks are now necessary.

1 Front fork oil change

Drain and replenish the damping fluid in the front forks. Refer to the relevant section in Chapter 4 for the relevant information.

2 Removal, inspection and relubrication of wheel bearings

Carry out the operations listed in the heading by following the procedure given in Chapter 5, Section 11 for the front wheel and Section 12 for the rear wheel.

3 Front and rear brake shoe inspection

On drum brake models both wheels should be removed and the brake shoes inspected for wear. Refer to Chapter 5 Sections, 9, 10 and 14 for the relevant details.

4 Steering head bearings

During the yearly service, the steering head bearings should be inspected and repacked with grease as described in Chapter 4 Section 6. To gain access to the bearings, the front forks must be removed as described in Section 2 and 3 of that Chapter.

5 Swinging arm bearings

The swinging arm should be removed from the machine, and the bearings re-packed with grease. See Chapter 4, Sections 9 and 10.

General maintenance adjustments

1 Clutch adjustment

The intervals at which the clutch should be adjusted will depend on the style of riding and the conditions under which the machine is used.

Adjust the clutch in two stages as follows:

Remove the clutch adjustment cover, which is retained by two screws. Loosen the cable adjuster screw locknut and turn the adjuster inwards fully, to give plenty of slack in the inner cable. Loosen the adjuster screw locknut in the casing and turn the screw clockwise until slight resistance is felt. Back off the screw about $\frac{1}{4}$ turn and tighten the locknut. The cover may be replaced.

Undo the cable adjuster screw at the handlebar lever, until there is approximately 2 – 3 mm (0.08 – 0.12 in) play measured between the inner face of the lever and the stock face. Finally, tighten the cable adjuster locknut.

2 Checking brake pad wear

Brake wear depends largely on the conditions in which the machine is ridden and at what speed. It is difficult therefore, to give precise inspection intervals, but it follows that pad wear should be checked more frequently on a hard ridden machine.

The condition of the brake pads can be determined with the pads in place in the caliper by viewing them through the inspection aperture in the caliper cover. The aperture is closed by a small plastic cap which may be hinged back to give access. Each pad has a red marked groove around the periphery. If it can be seen that one or both pads have worn down to or past the groove, the pads must be renewed as a set.

Pad removal can be accomplished without removing the front wheel, as follows:

Remove the single bolt which passes through the piston/cylinder casting into the caliper support bracket. It is upon this bolt that the casting slides. From the rear of the unit remove the single crosshead screw by passing the shank of a screwdriver through the wheel spokes. Grasp the piston/cylinder casting and lift it away, leaving the two pads in place on the support bracket either side of the disc.

To prevent the piston being expelled from the cylinder, in the event of the brake lever being applied inadvertently, place a wooden wedge between the piston and outer wall of the casing.

Lift each pad away from the disc and out of the support bracket. Note that the outer pad on the XS 360C model is fitted with an anti-chatter shim. This should be detached from the pad.

As stated above the pads must be renewed as a pair if either pad has worn down to the groove. The minimum pad thickness is 1.5 mm (0.06 in).

Replace the new pads by reversing the dismantling procedure. On XS 360C models, the anti-chatter spring must be fitted to the outer pad so that the stamped arrow mark is facing the direction of normal wheel travel. The outer faces of both pads should be smeared with a thin coating of silicon grease of the type supplied especially for disc brakes. Do not allow any grease to find its way onto the friction surfaces of the pad. It goes without saying that a pad contaminated with grease will not function efficiently.

Make sure that the brake pads are correctly located in the caliper and that the front wheel revolveves quite freely when reassembly is complete. Always check the brake action before taking the machine on the road.

Summary of routine maintenance, adjustments and capacities

Spark plugs	NGK BP-7ES (XS250 and 400), BP-6ES (XS360)
Spark plug gap	0.7 to 0.8 mm (0.028 to 0.032 in)
Contact breaker gap	0.3 to 0.4 mm (0.012 to 0.016 in)

Valve clearance (cold engine)

Inlet	0.08 to 0.12 mm (0.003 to 0.005 in)
Exhaust	0.16 to 0.20 mm (0.006 to 0.008 in)

Tyre pressures

	solo	pillion or at high speed
Front	26 psi (1.8 kg/cm²)	28 psi (2.0 kg/cm²)
Rear	28 psi (2.0 kg/cm²)	33 psi (2.3 kg/cm²)

Fuel tank capacity	11.0 litres (2.9/2.4 US/Imp gallons)

Engine oil capacity

With filter change	2.3 litres (5.0/4.0 US/Imp pints)
Without filter change	2.0 litres (4.2/3.5 US/Imp pints)

Front fork leg capacity	130 cc (4.4/3.7 US/Imp fl ozs)

Recommended lubricants

Component	Lubricant
Engine/gearbox unit	SAE 20W/50 engine oil
Telescopic forks	SAE 10W/30 engine oil, SAE 20 fork oil or ATF
Hydraulic front brake master cylinder	DOT 3 or SAE J1703 clutch and brake fluid
Control cables	Light engine oil
Grease nipples and wheel bearings	High melting point general purpose grease
Chain	Chain lubricant or graphited grease

Working conditions and tools

When a major overhaul is contemplated, it is important that a clean, well-lit working space is available, equipped with a workbench and vice, and with space for laying out or storing the dismantled assemblies in an orderly manner where they are unlikely to be disturbed. The use of a good workshop will give the satisfaction of work done in comfort and without haste, where there is little chance of the machine being dismantled and reassembled in anything other than clean surroundings. Unfortunately, these ideal working conditions are not always practicable and under these latter circumstances when improvisation is called for, extra care and time will be needed.

The other essential requirement is a comprehensive set of good quality tools. Quality is of prime importance since cheap tools will prove expensive in the long run if they slip or break when in use, causing personal injury or expensive damage to the component being worked on. A good quality tool will last a long time, and more than justify the cost.

For practically all tools, a tool factor is the best source since he will have a very comprehensive range compared with the average garage or accessory shop. Having said that, accessory shops often offer excellent quality tools at discount prices, so it pays to shop around. There are plenty of tools around at reasonable prices, but always aim to purchase items which meet the relevant national safety standards. If in doubt, seek the advice of the shop proprietor or manager before making a purchase.

The basis of any tool kit is a set of open-ended spanners, which can be used on almost any part of the machine to which there is reasonable access. A set of ring spanners makes a useful addition, since they can be used on nuts that are very tight or where access is restricted. Where the cost has to be kept within reasonable bounds, a compromise can be effected with a set of combination spanners – open-ended at one end and having a ring of the same size on the other end. Socket spanners may also be considered a good investment, a basic $3/8$ in or $1/2$ in drive kit comprising a ratchet handle and a small number of socket heads, if money is limited. Additional sockets can be purchased, as and when they are required. Provided they are slim in profile, sockets will reach nuts or bolts that are deeply recessed. When purchasing spanners of any kind, make sure the correct size standard is purchased. Almost all machines manufactured outside the UK and the USA have metric nuts and bolts, whilst those produced in Britain have BSF or BSW sizes. The standard used in USA is AF, which is also found on some of the later British machines. Others tools that should be included in the kit are a range of crosshead screwdrivers, a pair of pliers and a hammer.

When considering the purchase of tools, it should be remembered that by carrying out the work oneself, a large proportion of the normal repair cost, made up by labour charges, will be saved. The economy made on even a minor overhaul will go a long way towards the improvement of a toolkit.

In addition to the basic tool kit, certain additional tools can prove invaluable when they are close to hand, to help speed up a multitude of repetitive jobs. For example, an impact screwdriver will ease the removal of screws that have been tightened by a similar tool, during assembly, without a risk of damaging the screw heads. And, of course, it can be used again to retighten the screws, to ensure an oil or airtight seal results. Circlip pliers have their uses too, since gear pinions, shafts and similar components are frequently retained by circlips that are not too easily displaced by a screwdriver. There are two types of circlip pliers, one for internal and one for external circlips. They may also have straight or right-angled jaws.

One of the most useful of all tools is the torque wrench, a form of spanner that can be adjusted to slip when a measured amount of force is applied to any bolt or nut. Torque wrench settings are given in almost every modern workshop or service manual, where the extent to which a complex component, such as a cylinder head, can be tightened without fear of distortion or leakage. The tightening of bearing caps is yet another example. Overtightening will stretch or even break bolts, necessitating extra work to extract the broken portions.

As may be expected, the more sophisticated the machine, the greater is the number of tools likely to be required if it is to be kept in first class condition by the home mechanic. Unfortunately there are certain jobs which cannot be accomplished successfully without the correct equipment and although there is invariably a specialist who will undertake the work for a fee, the home mechanic will have to dig more deeply in his pocket for the purchase of similar equipment if he does not wish to employ the services of others. Here a word of caution is necessary, since some of these jobs are best left to the expert. Although an electrical multimeter of the AVO type will prove helpful in tracing electrical faults, in inexperienced hands it may irrevocably damage some of the electrical components if a test current is passed through them in the wrong direction. This can apply to the synchronisation of twin or multiple carburettors too, where a certain amount of expertise is needed when setting them up with vacuum gauges. These are, however, exceptions. Some instruments, such as a strobe lamp, are virtually essential when checking the timing of a machine powered by CDI ignition system. In short, do not purchase any of these special items unless you have the experience to use them correctly.

Although this manual shows how components can be removed and replaced without the use of special service tools (unless absolutely essential), it is worthwhile giving consideration to the purchase of the more commonly used tools if the machine is regarded as a long term purchase Whilst the alternative methods suggested will remove and replace parts without risk of damage, the use of the special tools recommended and sold by the manufacturer will invariably save time.

Chapter 1 Engine, clutch and gearbox

For information relating to later models, see Chapter 7

Contents

Specifications

Engine

	XS 250	XS 360	XS 400N
Type	Vertical parallel twin cylinder, overhead camshaft, four-stroke		
Bore	55 mm (2.1653 in)	66 mm (2.528 in)	69 mm (2.717 in)
Stroke		52.4 mm (2.063 in)	
Capacity	248 cc (15.13 cu in)	358 cc (21.85 cu in)	392 cc (23.92 cu in)
Compression ratio	9.6:1	9.6:1	9.2:1
bhp	27 @ 9,500 rpm	34 @ 8,500 rpm	—

Pistons and rings

Piston/cylinder clearance	0.030–0.050 mm (0.0012–0.0020 in)	
Piston oversizes available	+0.25 mm, +0.50 mm, +0.75 mm and +1.0 mm	
Piston ring end gap		
Top and 2nd ring	0.15–0.35 mm (0.006–0.014 in) 0.2–0.4 mm (0.008–0.016 in)	
Oil control ring	0.2–0.9 mm (0.008–0.035 in)	
Ring side clearance		
Top ring	0.04–0.08 mm (0.0016–0.0032 in)	
Second ring	0.03–0.07 mm (0.0012–0.0028 in)	

Cylinder barrel

Cylinder bore diameter	55.0 + 0.02 mm (2.17 + 0.0008 in)	66.0 + 0.02 mm (2.59 + 0.0008 in)	69.0 + 0.02 mm (2.72 + 0.0008 in)
Bore taper limit	0.05 mm (0.002 in)		
Ovality limit	0.01 mm (0.0004 in)		

Valves and springs

Valve clearance (cold)
Inlet ... ... 0.08–0.12 mm (0.003–0.005 in)
Exhaust ... ... 0.16–0.20 mm (0.006–0.008 in)
Valve stem diameter

Inlet ... ... $7.0 \begin{subarray}{l} -0.010 \\ -0.025 \end{subarray}$ mm $(0.275 \begin{subarray}{l} -0.0004) \\ -0.0009 \end{subarray}$ in$)$

Exhaust ... ... $7.0 \begin{subarray}{l} -0.030 \\ -0.045 \end{subarray}$ mm $(0.275 \begin{subarray}{l} -0.0012) \\ -0.0018 \end{subarray}$ in$)$

Stem/guide clearance
Inlet ... ... 0.010–0.037 mm (0.0004–0.0014 in)
Exhaust ... ... 0.030–0.057 mm (0.0012–0.0022 in)
Valve spring free length
Inner ... ... 39.3 mm (1.547 in)
Outer ... ... 42.8 mm (1.685 in)

Camshaft

Bearing inside diameter ... ... 23 + 0.021 mm (0.906 + 0.0082 in)

Journal outside diameter ... ... $23 \begin{subarray}{l} -0.020 \\ -0.033 \end{subarray}$ mm $(0.906 \begin{subarray}{l} -0.0008) \\ -0.0013 \end{subarray}$ in$)$

Bearing/journal ... ... 0.020–0.054 mm (0.0008–0.0020 in)
Cam height wear limit
XS250 and 400; inlet ... ... 39.38 mm (1.550 in)
              exhaust ... 39.42 mm (1.552 in)
XS360; inlet ... ... 38.70 mm (1.527 in)
           exhaust ... 38.74 mm (1.525 in)

Valve timing

	XS250 and 360	
Inlet opens	24° BTDC	30° BTDC
Inlet closes	60° ABDC	70° ABDC
Exhaust opens	58° BBDC	70° BBDC
Exhaust closes	26° ATDC	30° ATDC

Crankshaft

Big end clearance ... ... 0.021–0.045 mm (0.0008–0.0018 in)
Main bearing clearance ... ... 0.020–0.044 mm (0.0008–0.0016 in)
Connecting rod axial clearance ... 0.160–0.0264 mm (0.0063–0.0104 in)
Connecting rod deflection ... 0.30–0.50 mm (0.012–0.019 in)

Small end bore diameter ... ... $16 \begin{subarray}{l} +0.028 \\ +0.015 \end{subarray}$ mm $(1.614 \begin{subarray}{l} +0.0011 \text{ in}) \\ +0.0028 \end{subarray})$

Clutch

Type ... ... Wet, multi-plate
Number of plates
Friction 7
Plain 6
Number of springs 4
Friction plate thickness ... ... 3 mm (0.12 in)
Wear limit ... ... 2.7 mm (0.11 in)
Spring free length ... ... 34.6 mm (1.362 in)
Wear limit ... ... 33.6 mm (1.323 in)

Gearbox

Type						6-speed, constant mesh		
Ratios								
1st gear	...	...	...	...	2.500:1	2.500:1	2.500:1
2nd gear	...	...	...	...	1.777:1	1.777:1	1.777:1
3rd gear	...	...	...	...	1.380:1	1.380:1	1.380:1
4th gear	...	...	...	...	1.125:1	1.125:1	1.125:1
5th gear	...	...	...	...	0.961:1	0.961:1	0.961:1
6th gear	...	...	...	...	0.866:1	0.866:1	0.866:1
Secondary ratio		...	...	...	2.867:1	2.500:1	2.312:1

Torque wrench settings

Cylinder head					
6 mm bolts	...	...	...	...	0.8–1.2 kg/m (5.8–8.7 lbs/ft)
8 mm nuts (XS360 and 250)		...	2.0–2.4 kg/m (14.5–17.4 lb/ft)		
10 mm nuts (XS400)		...	...	3.3 kg/m (23.9 lb/ft)	
Primary drive pinion bolt		...	...	4.3 kg/m (31 lb/ft)	
Crankcase bolts					
6 mm	...	...	...	...	1.0 kg/m (7.2 lbs/ft)
8 mm	...	...	...	...	2.2 kg/m (16 lbs/ft)
Connecting rod cap bolts		...	...	3.3–3.8 kg/m (23.9–27.5 lb/ft)	

1 General description

The Yamaha XS250, 360 and 400 models use an engine unit which is identical in all major respects, there being differences only in the size of pistons and allied components relating to the engine capacity.

The engine is a vertical parallel twin with a 180° crankshaft, supported by three shell type main bearings. The 180° crankshaft has better inherent balance characteristics than the 360° crank more normally utilised in parallel twin engines, and dispenses with the necessity of a counter-balance assembly which is often fitted to modern vertical twins. The valve gear consists of an overhead camshaft driven by a chain from the crankshaft, passing through a tunnel between the two cylinders. The chain is automatically tensioned and rocker clearance is adjusted by means of adjuster screws on the rocker arms. All major engine casings are in light aluminium alloy, the cylinder block having dry steel liners. The engine/gearbox housings (crankcases) are split in a horizontal plane to facilitate dismantling and reassembly.

Engine lubrication is provided by a wet sump system ie; the oil reservoir is contained in the crankcase, not as is more usual, a separate oil tank mounted on the frame. Oil is picked up from the sump through a gauze oil trap and is fed under pressure by a trochoid oil pump through a corrugated paper filter to the working parts of the engine.

Primary drive is by double spur gears to a multi-plate wet clutch. A six-speed constant mesh gearbox then transmits drive via a roller chain to the rear wheel.

Two Mikuni constant vacuum carburettors and two independent exhaust systems are fitted.

An ac generator (alternator) mounted on the left-hand end of the crankshaft provides power to the electrical system. An electric starter motor is fitted to all but the XS 360-2D models, supplemented by a kickstart lever on the right-hand side of the machine.

2 Operations with engine/gearbox in the frame

It is not necessary to remove the engine unit from the frame for dismantling the following items:

1 *Cylinder head cover*
2 *Cylinder head, block and pistons*
3 *Right-hand and left-hand crankcase covers*
4 *Clutch assembly and oil pump*
5 *Alternator and starter motor*
6 *Replacement of the kickstart return spring*

If work on the crankshaft or the gearbox components is to be carried out it will be necessary to remove the engine from the frame in order that the crankcases may be separated. If a great deal of work is anticipated, it would be advantageous to remove the complete unit, thereby affording greater access.

3 Method of engine/gearbox removal

As mentioned previously the engine/gearbox is built in unit and it is necessary to remove the complete crankcase in order to gain access to the components. Separation is accomplished after the engine unit has been removed from the frame, and refitting cannot take place until the crankcase has been reassembled.

4 Removing the engine/gearbox unit

1 Place the machine on its centre stand, making sure that it is standing securely on firm ground.
2 Place a receptacle of more than 2·3 litres (5·0/4·0 US/Imp pints) below the engine and remove the filler cap from the primary chaincase and the drain plug from the base of the crankcase. A secondary drain plug is fitted in the crankcase left-hand wall, below the alternator casing. Oil drainage will be facilitated if the engine has been allowed to reach normal working temperature; the oil will be thinner and so flow more easily. Detach the oil filter chamber and oil filter element from the front of the crankcase by unscrewing the central bolt. Move the drain bath forwards to catch the small quantity of oil contained within the filter housing.
3 Raise the dualseat and disconnect the negative lead (–) from the battery, followed by the positive lead. This will isolate the electrical system and so prevent accidental shorting during subsequent dismantling. If it is anticipated that the machine is to be unused for an extended length of time, the battery should be removed for storage and given a refresher charge from an external source at approximately monthly intervals.
4 Turn the petrol tap to the On or Reserve position and disconnect the petrol feed pipe and the smaller gauge vacuum tube. Both tubes are secured on the unions by spring clips, the ears of which should be pinched together to release the tension. Drainage of the petrol tank for removal is not strictly necessary, although the reduction in weight will facilitate this operation. The petrol tank is supported at the front by two rubber buffers, which engage with a cup each side of the frame top tube, and is secured at the rear by a single bolt passing through a projecting lug. Remove the bolt and ease the petrol tank rearwards so that the buffers leave the cups. The tank may then be lifted upwards

away from the machine. Remove the two side covers from either side of the machine. The covers are a three point push fit in rubber grommets attached to the frame. Unscrew the screw clips which secure the carburettors to the inlet stubs, and the air hose unit to the carburettor mouths and the twin air filter boxes. Each air filter box may be pulled from position after removing the retaining strap, which is secured at the lower end by a single bolt and at the top by a hook arrangement. Disconnect the air hose breather tube from the breather cover on the crankcase, after displacing the spring securing clip. The air hose unit can now be pulled off the carburettor mouths and tilted backwards and downwards so that it clears the carburettors. Careful manipulation may be required as the clearance is limited. Pull the carburettors back out of the inlet stubs and lift them towards the right-hand side of the machine, as a unit.

5   Disconnect the throttle cable at the carburettor end. Lift the outer cable from the slotted abutment and displace the nipple and inner cable end from the throttle pulley between the two instruments.

6   Remove the tachometer drive cable from the front right of the camshaft cover. The cable is retained by a countersunk crosshead screw.

7   Disconnect the starter motor lead from the upper terminal on the starter solenoid switch. Pull the wire through, towards the engine, so that it will not get entangled when the engine is removed. The lead is secured to the machine at a number of points along its length by small lugs projecting from the rear mudguard (starter motor not fitted to XS360-20 model).

8   Disconnect the main electrical connections to the engine. The main alternator leads are connected at two block type multi-pin sockets behind the frame left-hand cover. The low tension leads should be disconnected from the ignition coil snap connectors above the cylinder head. The main earth lead from the battery to the engine is retained by a crankcase bolt to the rear of the gearbox. The lead may be disconnected by removing the bolt, or may be pulled through towards the engine and removed after lifting out the engine.

9   Undo the two nuts holding each exhaust clamp to the cylinder head and slide the clamps back. Note the split collets, and remove them. Slacken the bolt securing the two exhaust pipe/silencer joint clamps. The exhaust pipes may be pulled from position individually, leaving the silencers in place. Silencer removal is not necessary when removing the engine: each unit may be tipped forwards on the single mounting stud to clear the front of the engine. Prise out the exhaust port gaskets with a small screwdriver.

10   Remove the kickstart lever and the gear lever from their respective splined shafts. Both levers are retained by a pinch bolt which must be removed completely before the lever can be pulled off the shaft.

11   Remove the five screws which hold the final drive sprocket casing to the left-hand side of the engine. Pull the casing out until it is clear of the gearchange shaft. The clutch cable must be disconnected from the operating arm inside the case before the casing is completely free. Take care not to lose the small steel ball which lies between the inner face of the clutch lifting mechanism and the clutch pushrod, in the centre of the mainshaft.

12   Before removing the master link from the chain and running it off the sprockets, the gearbox sprocket nut should be loosened. Loosening this large nut after the engine has been removed can be accomplished but preventing the sprocket from rotating is more difficult. Bend down the tab washer before fitting the spanner and then apply the rear brake pedal to hold the sprocket. The sprocket may be pulled off the shaft with the chain in mesh, or after separation of the chain.

13   Remove the rear brake pedal from the splined pivot shaft after unscrewing completely the pinch bolt. Then detach both front footrests, each of which is held on two projecting stubs and retained by two bolts. Disconnect the stop lamp switch spring from the brake pivot arm and then lift the switch out of the holding bracket. Tuck the switch out of harm's way, to the right of the battery carrier.

14   The engine is held in position by two short bolts at the front and two long through bolts at the rear, the upper of which passes through two detachable mounting brackets. In addition, a head steady is fitted to a lug projecting from the camshaft cover. Commence engine removal by detaching the head steady plates. The plates are held by two bolts passing through frame lugs and a single bolt passing through the camshaft cover lug. Remove the upper rear engine bolt and the two mounting plates, together with the stop lamp switch bracket. The lower bolt may now be removed, lifting the engine if necessary and then allowing it to rest on the frame lugs. During engine removal the two left-hand exhaust port studs may foul the frame down tube. To prevent unnecessary damage to the paintwork, apply a thick covering of tape to the threads.

15   The engine can now be lifted out of the frame to the right-hand side. It is recommended that an assistant is present during engine removal to avoid the danger of dropping the heavy assembly.

4.4a Detach the fuel pipe at the tap union and ...

4.4b ... the tap vacuum pipe at the inlet stub

4.5a Pull air hose unit from carburettors and ...

4.5b ... pull carburettors out of the inlet stubs

4.5c Disconnect the throttle cable at the pulley

4.6 Tachometer cable is held by a single screw

4.7 Detach starter lead at solenoid upper terminal

4.8a Main alternator leads connected by block connectors

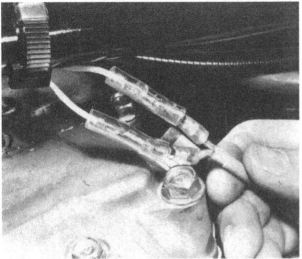

4.8b Disconnect the two separate contact breaker wires

4.9a Loosen silencer/exhaust pipe clamps and ...

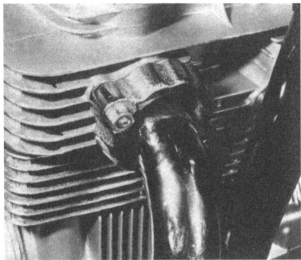

4.9b ... free the exhaust pipe flanges

4.9c Slacken silencer mounting nut and tip silencer forwards

4.10 Kickstart and gear levers held by a pinch bolt

4.11 Detach clutch cable at operating mechanism

4.12 Bend down washer and loosen nut before chain removal

4.14a Detach the head steady, held by three bolts

14.4b Engine is secured at front by two short bolts and ...

14.4c ... held at lower rear by a single long bolt

4.14d Detach upper bolt and also mounting plates

4.15 Lift out engine towards right-hand side

5   Dismantling the engine/gearbox: general

1   Before undertaking any work on the engine unit it should be thoroughly cleaned with a proprietary de-greaser to remove the grease and grit which accumulates. If this precaution is not followed then there is a very likely possibility of foreign particles entering the dismantled engine and components. **Note:** Make sure to block up the inlet and exhaust ports when washing down, to stop water entering the engine.
2   If the engine has not been stripped before, it is strongly advisable to have an impact screwdriver complete with a set of crosshead bits available. If an impact screwdriver is unavailable, it is sometimes possible to use a crosshead screwdriver fitted with a T handle. Whenever an attempt is made to slacken a screw, assume initially that it is very tight and proceed accordingly. Avoid a half-hearted attempt at loosening as the screwdriver may slip, causing damage to the head and so prevent subsequent successful removal.
3   Do not use force to remove any parts unless specific mention is made in the manual. If any component is difficult to

remove, first check that everything has been dismantled and/or loosened in the correct sequence.

6   Dismantling the engine/gearbox: removing the contact breaker assembly and automatic timing unit

1   Remove the two screws which hold the contact breaker cover plate to the left-hand side of the cylinder head cover. The twin contact breaker assembly plate is retained in position by two clamping screws and large plain washers; the screws do not pass through the base plate but clamp it in place. Before loosening the screws, scribe a line across the plate and onto the edge of the casing. This will aid correct positioning on reassembly.
2   The automatic timing unit (ATU) is positioned behind the contact breaker base plate. Remove the centre bolt and pull the ATU off the end of the camshaft. Note that a dowel pin fitted in the end of the camshaft locates with the ATU and ensures that the ATU/camshaft relationship is always correct.

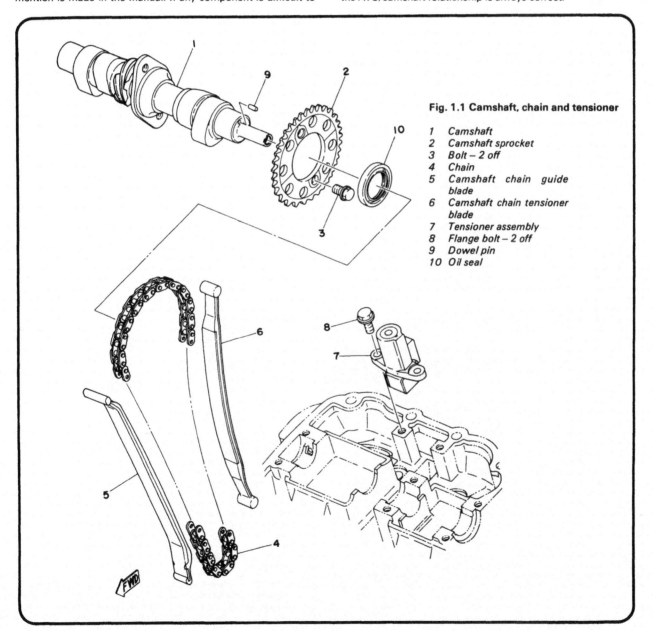

Fig. 1.1 Camshaft, chain and tensioner

1   Camshaft
2   Camshaft sprocket
3   Bolt – 2 off
4   Chain
5   Camshaft chain guide blade
6   Camshaft chain tensioner blade
7   Tensioner assembly
8   Flange bolt – 2 off
9   Dowel pin
10 Oil seal

6.1 Remove the contact breaker assembly complete

6.2 The ATU is retained by a single bolt

7 Dismantling the engine/gearbox: removing the camshaft cover, and cylinder head

1   Position the engine on the workbench, resting it securely on the crankcase base with a wood block placed between the front mounting lugs and the bench.

2   The camshaft cover is retained by fourteen bolts which should be slackened in an even and diagonal sequence. The camshaft cover is under uneven stress from those valves and springs on partial or complete lift and therefore the holding down bolts must be unscrewed evenly to avoid distortion of the cover.

3   With all the bolts removed, lift the camshaft cover away. Note that no gasket or sealing 'O' ring is fitted between the mating surfaces of the camshaft cover and cylinder head.

4   Remove the two bolts which retain the cam chain guide holder. Lift off the holder and pull out the chain tensioner blade and also the chain guide blade, from the forward edge of the cam chain tunnel.

5   In order to gain access to the two cam chain sprocket bolts, it will be necessary to rotate the crankshaft through about 90°. This is best accomplished after the alternator cover has been removed, when a spanner can be fitted on the alternator rotor

bolt. The alternator cover and gasket are retained by three countersunk screws. Remove the two sprocket bolts and pull the sprocket towards the left-hand end of the camshaft. This will bring the sprocket off the locating boss and allow the cam chain to be lifted off the sprocket.

6   The camshaft and sprocket can now be eased out of position in the cylinder head. To prevent the cam chain falling down into the crankcase, place a screwdriver or short rod through the chain so that it rests across the camshaft half-bearings. Although this is not so important if the engine is being dismantled completely, it may prove difficult to retrieve the cam chain if only the top end of the engine is being overhauled.

7   The cylinder head is retained by eight acorn nuts and two bolts. First loosen the two bolts which are fitted one each side of the head, adjacent to the spark plug holes. Loosen the eight acorn nuts in the reverse order given in Fig. 1.14 which accompanies Section 42. This procedure will prevent distortion of the complex aluminium structure.

8   The cylinder head can now be eased up, off the holding down studs. It may be necessary to use a soft-nosed mallet to aid removal. Do not strike the cylinder head fins except where they are well supported by the porting. **Do not** use levers to lift the head or broken or bent fins will result. When the head is clear of the cylinder block, slide another screwdriver between the cam chain and remove the screwdriver from the top of the cylinder head.

9   Pull the three hollow dowel pins from the top of the cylinder block, noting the rubber seal on the rear right-hand dowel. Remove and discard the cylinder head gasket.

7.6 Do not allow chain to fall into crankcase

8 Dismantling the engine/gearbox: removing the cylinder block and pistons

1   After removing the cylinder head, the block is free to be lifted off. If it is stuck by the gasket, careful use of a hide hammer may be required. When lifting the block off the connecting rods, the pistons should be supported to stop them falling against the edge of the crankcase mouth, which will damage them. After raising the cylinder block from the base gasket, but before the piston rings leave the cylinder bore spigots, pad the crankcase mouths with clean rag. This will prevent pieces of broken piston ring, or gudgeon pin circlips from falling into the crankcase. This is particularly important when a top-end overhaul only is being carried out. After removing the cylinder block, pull out the three hollow dowels and the O ring.

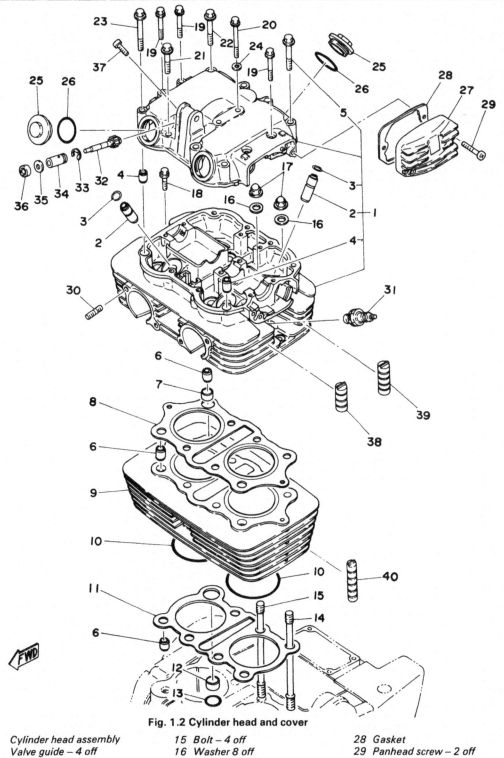

Fig. 1.2 Cylinder head and cover

1	Cylinder head assembly	15	Bolt – 4 off	28	Gasket
2	Valve guide – 4 off	16	Washer 8 off	29	Panhead screw – 2 off
3	Circlip – 4 off	17	Nut – 8 off	30	Stud – 4 off
4	Dowel pin – 2 off	18	Bolt – 2 off	31	Spark plug – 2 off
5	Bolt	19	Bolt – 4 off	32	Gear pinion (15T)
6	Dowel pin – 5 off	20	Bolt – 2 off	33	Circlip
7	Seal	21	Bolt	34	Housing
8	Gasket	22	Bolt – 4 off	35	Washer
9	Cylinder block	23	Bolt – 2 off	36	Oil seal
10	'O' ring – 2 off	24	Washer – 2 off	37	Screw
11	Gasket	25	Camshaft cover – 4 off	38	Damper rubber
12	Hollow dowel	26	'O' ring – 4 off	39	Damper rubber
13	'O' ring	27	Contact breaker cover	40	Damper rubber
14	Bolt – 4 off				

2   Remove both the circlips from each piston and drift out the gudgeon pin using a correct diameter drift. Support both the piston and the connecting rod during this operation to prevent any undue sideways strain. Do not completely remove the gudgeon pin but leave it in its respective piston boss. After removing each piston, mark it on the inside of the skirt so it can be replaced in its correct bore on subsequent reassembly. If the gudgeon pin is a tight fit in the piston, the piston should be gently heated by placing a rag soaked in boiling water on the crown. Do not use excessive force to remove a gudgeon pin because of the possibility of bending a connecting rod.

3   There are no separate small end bearings, the gudgeon pin bears directly onto the bearing surface of the connecting rod.

4   The piston rings can be removed by gently opening the gap and lifting them one at a time off the piston. Great care must be exercised since the rings are very brittle and thus easy to break. If it is intended to use the rings again note the order in which they were removed and also which was the top and bottom surface. If difficulty is encountered in removing the piston rings, three thin strips of tin, cut from an oil can, may be used to aid removal (see illustration).

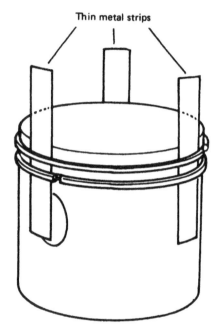

Fig. 1.3 Freeing gummed piston rings

9   Dismantling the engine/gearbox: removing the alternator rotor and the starter motor

1   Remove the alternator outer cover (if this has not already been done) to aid crankshaft rotation, as described in Section 7.5.

2   The inner cover, which houses the alternator field coils and encases the starter motor drive chain, is retained by socket bolts. Remove the inner cover and gasket after disconnecting the neutral warning switch lead and oil pressure warning switch lead from their respective switches. Each lead terminal is retained by a small screw. Displace the two alternator cables from the securing clips to free the case completely.

3   Remove the chromed cover which encloses the starter motor chamber and which is retained by two bolts.

4   Unscrew the two crosshead screws at the rear of the starter motor. Withdraw the starter motor whilst holding the starter motor sprocket. The starter motor sprocket is not retained on

the motor shaft splines by any fixing system; it is a sliding fit. It may be necessary to lightly tap the starter motor out of the crankcase with a hide mallet, since it is often a tight fit in the housing. **Warning:** Do not hit the starter motor on the splined shaft as this will damage the planetary reduction gear. Lift the starter motor out of the chamber together with the main lead which is retained on the terminal on the underside of the motor by a nut.

5   The starter motor chain and sprocket are now free and can be lifted away from the large driven sprocket.

6   Remove the driven sprocket guide plate which is retained by a single bolt.

7   To remove the alternator rotor retaining bolt it is necessary to stop the engine turning. This is best achieved by placing a metal bar through one of the little end bosses. Rest this bar on pieces of wood placed on top of the crankcase mouth. On no account must the metal bar be allowed to bear directly down onto the gasket face; if it does it will most certainly damage the jointing surface and cause oil leaks.

8   After jamming the engine to stop movement, remove the retaining bolt and washer from the alternator rotor. This will uncover an internal thread in the centre boss of the rotor. This thread facilitates the use of the Yamaha extractor tool.

9   Remove the rotor from its tapered shaft. If the special tool is unavailable a two or three jaw puller can be used instead. Note there are cutaways in the rotor which can be utilised by the puller's legs. Do not forget to remove the Woodruff key. Replace the centre bolt loosely during extraction with a puller, to prevent damage to the internal thread.

10   The large sprocket is a push fit over the crankshaft and can be lifted clear after removal of the small retention tongue held by a single bolt.

11   The starter motor free running clutch is contained at the rear of the rotor. For further details refer to Chapter 6, Section 16.

10   Dismantling the engine/gearbox: removing the neutral indicator switch

1   The neutral indicator switch is retained on the gearbox left-hand wall by three countersunk screws. Slacken the screws evenly and remove them, followed by the switch body. Make a scribe mark on the body to aid correct repositioning on subsequent reassembly.

2   Removal of the tiny contact brush and spring from the hole in the gearchange drum end is not necessary unless the contact is worn or the spring weakened.

9.2a Detach oil pressure switch lead and ...

9.2b ... neutral indicator switch lead

9.4 Lift starter motor out of the chamber

9.9 Sprocket puller makes good substitute for Yamaha extractor

2   Unscrew the four bolts in the clutch pressure plate, and lift off the plate complete with bolts, springs and washers. Remove the clutch operating thrust piece and the clutch operating push rod. Note that a small steel ball is fitted between these components. Displace the ball, using a length of stiff wire, and store it in a safe place. Lift out the clutch plates either one at a time or as a sandwich, and note their sequence to aid reassembly.
3   Remove the circlip from position on the mainshaft end and pull off the clutch centre boss followed by the clutch outer drum.
4   The primary drive pinion is a tight push fit on the crankshaft end, where it is retained by a single bolt and located by a Woodruff key. Lock the crankshaft with a close fitting bar through one small end eye. Using the previously described method remove the centre bolt and then draw the pinion off the shaft. If necessary, place two levers between the casing wall and the rear face of the pinion, to facilitate removal. Care must be taken not to damage the casing. Prise the Woodruff key from the recess in the shaft.

11 Dismantling the engine/gearbox: removing the primary drive cover and the oil pump

1   Loosen evenly and remove the screws which hold the primary drive cover in place. The cover is located on two dowel pins and may require tapping from position with a rawhide mallet.
2   The oil pump is secured by three large socket bolts and is located by a single dowel. Remove the bolts and lift the unit away from the casing, complete with the idler gear. The idler gear is a free sliding fit on the stub shaft.

12 Dismantling the engine/gearbox: removing the kickstart assembly, the clutch and primary drive pinion

1   Grasp the outer turned end of the kickstart return spring with a pair of pliers and ease it off the anchor lug projecting from the crankcase. Allow the tension on the spring to be released in a controlled manner. The kickstart assembly may be withdrawn from the casing as a complete unit and placed to one side, for examination at a later stage.

11.2 Remove the three socket bolts holding the oil pump

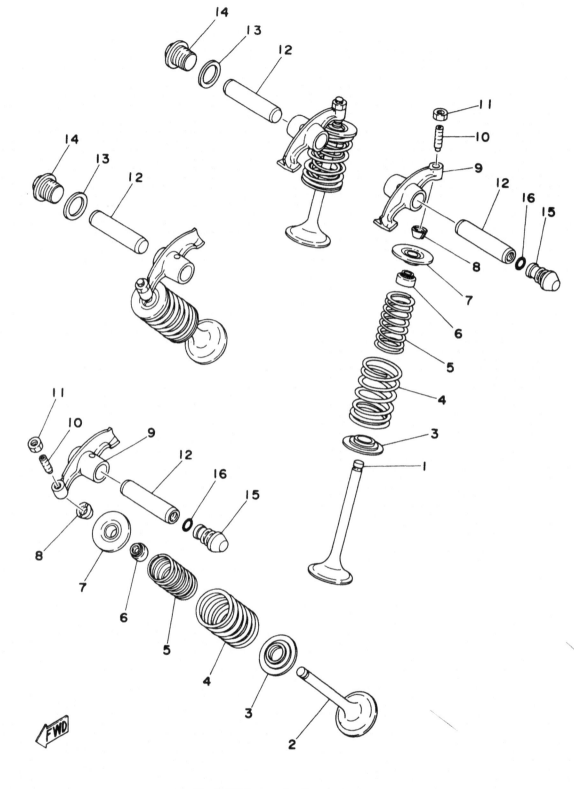

Fig. 1.4 Valves and rocker arms

1 Inlet valve – 2 off	7 Valve collar – 4 off	12 Rocker arm spindle – 4 off
2 Exhaust valve – 2 off	8 Split collet – 8 off	13 Washer – 2 off
3 Valve spring seat – 4 off	9 Valve rocker arm – 4 off	14 Plug – 2 off
4 Outer spring– 4 off	10 Valve clearance adjusting screw	15 Plug – 2 off
5 Inner spring – 4 off	– 4 off	16 'O' ring – 2 off
6 Valve stem oil seal – 4 off	11 Nut – 4 off	

13 Dismantling the engine/gearbox: removing the gearchange mechanism

1   Pull the rubber boot from the left-hand end of the gearchange shaft and prise off the 'E' clip, removing it, together with the washer. Withdraw the shaft from the right-hand side of the engine and lift the small shouldered pivot bush from the stub shaft projecting from the gear selector arm. The arm is retained on the extended front gear selector fork rod by an 'E' clip. Remove the 'E' clip and hold the two pawl arms out, away from the change drum pins, to allow removal of the arm.

2   Loosen and remove the four screws which hold the two small plates either side of the change drum end. These screws are very tight and will require a concerted attempt to remove them. Special care should be taken when loosening the dome headed screws as the heads are easily damaged.

14 Dismantling the engine/gearbox: separating the crankcase halves

1   Invert the engine so that the lower crankcase faces upwards and the upper crankcase rests securely on the cylinder holding down studs and the rear portion of the upper gearbox casing.

2   Loosen the crankcase holding bolts evenly by reversing the sequence given in Fig. 1.11 accompanying Section 35 of this Chapter. Each holding bolt should be loosened initially about ¼ turn and then the sequence repeated.

Invert the crankcase once more and remove the upper bolts following the correct sequence. Double check that all the bolts have been removed before attempting to separate the cases. A rawhide mallet may be used to separate the crankcase halves; the use of levers should be avoided as the mating surfaces are easily damaged, leading to oil loss once the engine is in service again.

3   The upper crankcase half should be lifted away, leaving the crankshaft and gearbox components in place in the casing lower half.

The three main bearing shells in the upper casing should be left in place at this stage. If they fall out when lifting the casing, restore them to their original positions to avoid confusion.

15 Dismantling the engine/gearbox: removing the crank-shaft and gearshaft assemblies

1   The crankshaft assembly can be lifted out of position, and the cam chain removed as soon as the crankcases have been separated.

2   Lift out the layshaft as a complete unit and then remove the mainshaft. Displace the half clips which locate the bearings axially and remove from the mainshaft the clutch pushrod oil seal.

3   The three main assemblies should be placed to one side for later examination and dismantling.

16 Dismantling the engine/gearbox: removing the gear selector mechanism

1   Invert the crankcase lower half and using a large screwdriver remove the change drum detent housing bolt, spring and scrolled plunger. Right the casing to gain access to the selector fork rods.

2   Prise the 'E' clip from the rear selector fork rod, where it locates in the annular groove adjacent to the gearbox left-hand wall. Using a suitable drift, drive out gently the selector fork rod towards the left-hand side. This will displace the blind grommet in the gearbox wall. Pull the rod out completely to free the two selector forks. It is suggested that each of the four forks be marked clearly, to aid replacement. The guide pins are a light

push fit in the forks and should be removed to avoid loss. Drive out the remaining selector fork rod in a similar manner.

3   Pull the gearchange drum out of its bearing as far to the primary drive side as possible. Prise off the circlip on the drum end and remove the drum stopper cam plate. Remove the small peg from the drum end. The change drum is now free to be withdrawn completely.

17 Dismantling the engine/gearbox: removing the oil baffle plate and the oil strainer screen and cover

1   A detachable baffle plate is fitted within the crankcase lower half, which may be removed to facilitate casing cleaning. The baffle plate is retained under its own tension by a transverse rod, which passes over the top of the plate and locates in apertures in the casing walls. Removal of the plate may be effected by pressing downwards firmly on the plate and withdrawing the rod.

2   A filter screen is fitted in the underside of the crankcase, enclosed by a plate held by six bolts. Remove the bolts and break the seal between the cover and gasket, using a rawhide mallet. Lift the screen from position

18 Examination and renovation: general

1   Before examining the parts of the dismantled engine unit for wear, it is essential that they should be cleaned thoroughly. Use a paraffin/petrol mix to remove all traces of old oil and sludge which may have accumulated within the engine. Neat paraffin, although an admirable solvent, has a low evaporation rate. As a consequence of this, difficulty is often encountered in drying cleaned components.

2   Examine the crankcase castings for cracks or other signs of damage. If a crack is discovered, it will require professional repair.

3   Examine carefully each part to determine the extent of wear, checking with the tolerance figures listed in the Specifications section of this Chapter or accompanying the text. A vernier gauge or external micrometer will be required for determining external dimensions. The running clearance between two working surfaces such as the piston and cylinder bore can often be checked using ordinary feeler gauges. Checking internal dimensions such as the diameter of a shell bearing or cylinder bore requires the use of an internal micrometer, the size of which may vary depending on the component concerned. After some experience has been gained, it is often possible to determine by eye the extent of wear, and hence the need for renewal, without resetting the direct measurement. If there is any question of doubt, play safe and renew.

4   Use a clean, lint-free rag for cleaning and drying the various components. This will obviate the risk of small particles obstructing the internal oilways, causing the lubrication system to fail.

19 Examination and renovation: main bearings and big-end bearings

1   The XS series twins are fitted with shell type bearings on the crankshaft and the big-end assemblies.

2   Bearing shells are relatively inexpensive and it is prudent to renew the entire set of main bearing shells when the engine is dismantled completely, especially in view of the amount of work which will be necessary at a later date if any of the bearings fail. Always renew the three sets of main bearings together. Replacement bearing shells are supplied on a selective fit basis (ie; bearings are selected for correct tolerance to fit the original journal diameter and the diameter of the bearing housing). Each bearing is selected by subtracting the crankshaft journal number from the crankcase housing number, and referring to the following selection table.

Housing No.	Journal No.	Shell No.
3	1	1 (Blue)
4	2	2 (Black)
5	—	3 (Brown)
—	—	4 (Green)

The identification numbers for the crankcase and journals are found on the crankcase upper half and the crankshaft webs respectively as shown in the accompanying illustrations. In practice, provided that the correct numbers are quoted, the supplier of the replacement bearings will be in a position to make the correct selection.

3   Wear is usually evident in the form of scuffing or score marks in the bearing surface. It is not possible to polish these marks out in view of the very soft nature of the bearing surface and the increased clearance that will result. If wear of this nature is detected, the crankshaft must be checked for ovality as described in the following Section.

4   Failure of the big-end bearings is invariably accompanied by a pronounced knock within the crankcase. The knock will become progressively worse and vibration will also be experienced. It is essential that bearing failure is attended to without delay because if the engine is used in this condition there is a risk of breaking a connecting rod or even the crankshaft, causing more extensive damage.

5   Some indication of big-end wear may be deduced by measuring the side-to-side movement of the connecting rod at the small-end eye. The connecting rod should be moved in line with the crankshaft. A deflection of 0·50 mm (0·019 in) or more indicates the need for bearing renewal. Check the axial float of the connecting rod between the big-end eye and the crankshaft checks, using a feeler gauge. A clearance of more than 0·264 mm (0·0104 in) indicates a worn crankshaft or connecting rod.

Before the big-end bearings can be examined, the bearing caps must be removed from each connecting rod. Each cap is retained by two high tensile bolts. Before removal, mark each cap in relation to its connecting rod so that it may be replaced correctly. As with the main bearings, wear will be evident as scuffing or scoring and the bearing shells must be replaced as two (2) complete sets.

Replacement big-end bearings are selected in a similar manner to that adopted for main bearing renewal, by referring to the same table. The housing numbers and journal size numbers will be found on the connecting rods and the extreme right-hand web on the crankshaft. Of the two numbers stamped side by side on the web, the first corresponds to the left-hand journal, and the second to the right-hand journal.

6   Both main bearing and big-end bearing shells are colour coded, as stated in the table. Check the colour marks to be doubly sure of the correct selection.

19.3a Check bearing shells for scoring or wear

19.3b Main bearing shells are colour marked to aid identification

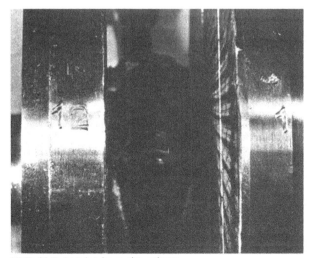

19.3c Main bearing journal numbers

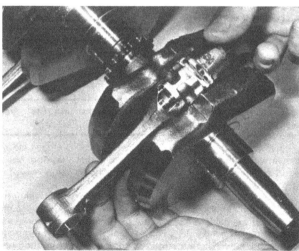

19.5a Separate the big-end bearing caps from connecting rods

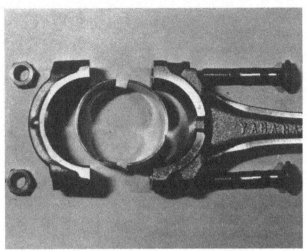

19.5b Big-end/connecting rod assembly: general view

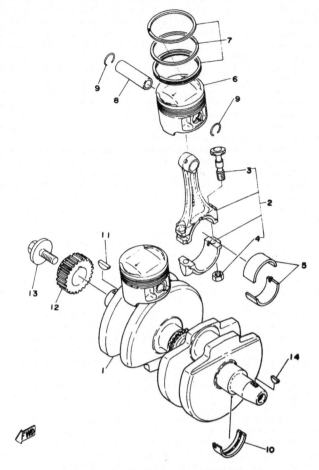

20 Examination and renovation: crankshaft assembly

1 If wear has necessitated the renewal of the big-end and/or main bearing shells, the crankshaft should be checked with a micrometer to verify whether ovality has occurred. If the reading on any one journal varies by more than 0·002 in (0·06 mm) the crankshaft should be renewed.

2 Mount the crankshaft by supporting both ends on V blocks or between centres on a lathe and check the run-out at the centre main bearing surfaces by means of a dial gauge. The run-out will be half that of the gauge reading indicated. The correct run-out as standard is under 0·02 mm (0·0008 in) and if it exceeds this the crankshaft should be renewed.

3 If scoring or flaking of the crankshaft is evident, renewal is the only remedy. Regrinding of the journals is not possible and no oversize bearing shells are available.

4 When refitting the connecting rods and shell bearings, note that under no circumstances should the shells be adjusted with a shim, 'scraped in' or the fit 'corrected' by filing the connecting rod and bearing cap or by applying emery cloth to the bearing surface. Treatment such as this will end in disaster; if the bearing fit is not good, the parts concerned have not been assembled correctly. This advice also applies to the main bearing shells. Use new big-end bolts too – the originals may have stretched and weakened.

Fig. 1.5 Crankshaft and piston assembly

1 Crankshaft	7 Piston ring set – 2 off
2 Connecting rod assembly – 2 off	8 Gudgeon pin – 2 off
	9 Circlip – 4 off
3 Connecting rod bolt – 4 off	10 Main bearing – 6 off
	11 Woodruff key
4 Nut – 4 off	12 Primary drive pinion (24T)
5 Big end bearing shell – 4 off	13 Bolt
	14 Woodruff key
6 Piston – 2 of	

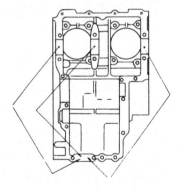

Main bearing housing numbers

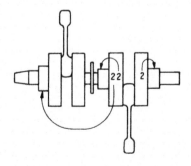

Main bearing journal numbers

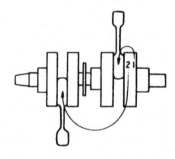

Big-end bearing journal numbers

Fig. 1.6 Shell bearing selection number locations

5   Oil the bearing surfaces before reassembly takes place and make sure the tags of the bearing shells are located correctly. Fit each connecting rod and cap so that the two tangs are to the rear of the engine and the mark Yamaha on each rod is facing towards the left-hand side of the engine. After the initial tightening of the connecting rod nuts, check that each connecting rod revolves freely, then tighten to a torque setting of 3·3 – 3·7 kg/m (24 – 27 ft/lbs). Check again that the bearing is quite free.

21 Examination and renovation: connecting rods

1   It is unlikely that either of the connecting rods will bend during normal usage, unless an unusual occurrence such as a dropped valve has caused the engine to lock. Carelessness when removing a tight gudgeon pin can also give rise to a similar problem. It is not advisable to straighten a bent connecting rod; renewal is the only satisfactory solution.
2   The small end eye of the connecting rod is unbushed and it will be necessary to renew the connecting rod if the gudgeon pin becomes a slack fit. Always check that the oil hole in the small end eye is not blocked since if the oil supply is cut off, the bearing surfaces will wear very rapidly.

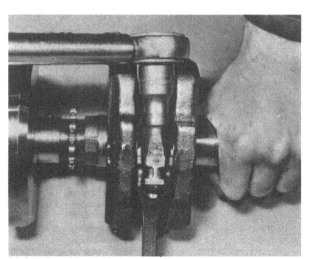

20.6 Tighten big-end bolts to correct torque

22 Examination and renovation: cylinder block

1   Cylinder bore wear is usually accompanied by excessive oil consumption and blue smoke from the exhaust. On examination of the top of the bores there will be a pronounced ridge at the limit of the travel of the top piston ring if wear has occurred. Do not mistake a ridge for a carbon deposit which can be scraped off.
2   Measure the bore, with an internal micrometer, just below the ridge (or at the limit of travel of the top piston ring) between the front and the rear of the barrel, where most wear occurs. Also measure the bore at the bottom of the cylinder. A comparison of the two results will give the cylinder bore wear.
    If no internal micrometer is available, place a piston ring squarely in the top of the bore, below the ridge, and measure the end gap with a feeler gauge. Repeat at the bottom of the bore. Subtract this later figure from the first and divide by three (more accurately 22/7). This will give the approximate wear. The bottom of the skirt of the piston can be used to position the ring squarely in the bore. Compare the results obtained with the service limits.
3   Check the bores for score marks; any deep damage will necessitate a rebore.

20.4 Ensure that bearing shell tags locate correctly

4   Check that the cooling fins are not clogged with oil and dirt otherwise overheating will occur. The fins can be cleaned by judicious use of a small screwdriver and/or a wire brush.

23 Examination and renovation: pistons and piston rings

1   If a rebore is required, due to the condition of the bores, new oversize pistons and rings will have to be obtained.
2   Remove the carbon from the piston crowns using a soft scraper (an old piece of aluminium bar which has been sharpened for the purpose makes an ideal scraper) and finally finish with metal polish. Emery cloth is not recommended but wire wool and oil can be successfully employed.
3   Wear is usually evident on the thrust side of the pistons and is noticeable as vertical streaks and/or score marks. If the score marks are not severe they may be smoothed out by light filing with a needle or swiss file.
4   The piston ring grooves may have become enlarged. Check them with a feeler gauge and compare the reading with the service limits.
5   Piston ring wear is difficult to estimate due to the variability in the end gap setting when the engine was last assembled. However, there is a service limit and the rings can be checked by placing them squarely in the bore and measuring the end gap with a feeler gauge. If the gap is greater than the maximum allowed, then renew the rings. However, if the gap is within the prescribed limits it does not positively show that excessive wear has not taken place. Therefore if in doubt or if a high mileage has been covered, renew the piston rings as a precaution.

24 Examination and renovation: cylinder head

1   The cylinder head, which was removed at an earlier stage, now contains only the valves, valve springs and associated components.
2   Remove the carbon from the head with the valves still in position, using the same technique as in the previous Section. It is advisable to put two old spark plugs in the sparking plug holes so that the threads do not become either damaged or blocked.
3   A valve spring compressor is required to remove the valves. On compressing the spring the valve collets can be lifted out and then the pressure released. This will free both the valve, valve spring collar and valve springs. Note where the valves come from so that they can be replaced in their respective seats. The inlet valve has the larger diameter head. Remove and

discard the valve oil seals which are a push fit on the valve guides. The seals should be renewed as a matter of course. The spring lower seat can now be lifted off.

4 Finish cleaning the head and valves and remove any dirt from between the head fins, to stop overheating.

5 Check that the valve guides are free from carbon, particularly the exhaust.

6 Visually inspect the valves for wear on the stem and/or the collet area. If excessive wear is evident or in doubt, replace the valves. Measure the valve stem in several places with a micrometer and compare readings with the service limit. If the valve is in good condition except for some deep pits in the seat, the valve can probably be refaced at a garage. Also check that the valve stem is straight by rolling the valve on a flat plate whereby any irregularities will be noticed.

7 Measure the internal diameter of the valve guides and compare readings with the service limits. Renew if necessary. If replacement is required the head should be uniformly heated to 100°C (212°F) in an oven (a kitchen oven) and the old guide drifted out with a correct sized drift. The new guide should then be immediately drifted in whilst the head is still hot. Do not omit the small circlip which is fitted to the guide. On cooling, the guide should be reamed out to the correct size using a copious amount of oil. If the guides have had to be replaced, the valve seats will also need recutting.

8 Inspect the valve seats and if they are deeply pitted, or new guides have been fitted, they will have to be recut at a garage. Both inlet and exhaust valve seats are cut to the same angle, see Specifications.

9 If the pitting of either the valves or valve seats is only slight, or if new valves have been fitted, the valves will require grinding in.

10 Valve grinding is basically a simple task – smear a little medium or fine grinding compound (carborundum paste), depending on the amount of pitting, onto the valve seat. Oil the valve guide and place the valve into position. Rotate it backwards and forwards using a suction valve grinding tool. Occasionally the valve should be lifted and replaced in a different position on the valve seat so as to grind in the whole seat evenly. Repeat this operation, using fine grinding paste, if this has not been used previously. Be careful not to get any grinding paste on the valve stem or in the valve guide. The valve is ground in when there is a complete and unbroken ring of light grey matt finish on both the valve and seat. Afterwards, throughly wash off the grinding paste to prevent damage that will be caused by its abrasive action. Do not rèsort to excessive grinding as this will pocket the valve in the seat.

11 Check the valve collets for wear or chipping and renew them if necessary.

12 The valve springs should be checked by measuring their free length and comparing the measurements with the service limits. Also check that the springs are not starting to crack or have worn badly by rubbing against each other. Replace them all if in any doubt.

13 Check that the cylinder head to block gasket face is in good condition and that the head is not warped. To check for warping use engineers blue on a surface plate, or a piece of plate glass. Failing this, check in several places, using a straight edge. Small high spots or slight warpage can be corrected by hand, with a flat oil stone.

25 Examination and renovation: camshaft, camshaft sprockets, chain and tensioner

1 The cam lobes should have a smooth surface, with no indentations. It is unusual to find wear unless a lubrication fault has occurred. If scuffing is apparent but is not serious, it should be stoned out with a carborundum stone and oil. If scuffing on the cam lobes is sufficient to alter the cam shape after stoning has been carried out, then the camshaft will need renewing. Measure the overall height of each cam from the base circle to the tip of the lobe. If the measurement is less than the service limit on any one cam, the camshaft should be renewed.

2 The camshaft does not have any bearings as such but rotates on the camshaft journals on bearing ways machined into the cylinder head and the camshaft cover. Check the condition of the bearing surfaces. It is unlikely that wear will have taken place on the cylinder head or camshaft cover, unless lubrication failure has occurred.

3 If wear has developed in the bearing surfaces as a whole the cylinder head, camshaft cover and camshaft must be renewed.

4 Check the camshaft sprocket for missing and/or chipped teeth. If side wear is apparent, it is probably due to worn camshaft bearings or incorrect endfloat. The lower sprocket on the crankshaft should also be checked. This component is an integral part of the crankshaft, and as such if wear occurs the crankshaft must be renewed. Fortunately this state of affairs rarely occurs because the crankshaft sprocket operates in ideal conditions.

5 Inspect the camshaft chain for wear, cracked links or missing rollers. An indication of wear is given by bending the chain sideways, across the line of the rollers; if a pronounced curve is formed, the chain is worn and should be renewed. Camshaft chains do not generally wear until considerable service has been seen, as they function in an ideal environment, always correctly lubricated.

6 The chain tensioner blade and the chain guide blade are made of plastic and will wear eventually. Light grooves in the blades can be overlooked, but if heavier wear has taken place, the blade should be renewed. The chain tensioner unit is unlikely to suffer wear or damage. Check that the plunger moves in or out smoothly after depressing the spring load pawl.

26 Examination and renovation: rocker arms and pins and tachometer driveshaft

1 Visually inspect the rocker arms for cracks and check the tips for excessive wear.

2 Check the rocker arm pins for wear and see that there is not an excessive clearance between the pin and arm. If there is too much clearance there will be a general rattle or clatter from inside the cambox. Inspect the tappet adjuster screw heads for hammering or flaking, renewing where necessary. The left-hand rocker pins are retained by rubber plugs which in turn are secured by the contact breaker plate. The right-hand pins are secured by screw plugs. Each pin has an internal thread in the outer end into which may be screwed a suitable bolt, to aid extraction.

3 The tachometer driveshaft is fitted to the camshaft cover where it is driven by a scroll gear integral with the camshaft. The shaft is unlikely to wear unless lubrication failure occurs. To remove the shaft and housing displace the 'E' clip from inside the cover. The housing may then be drifted out, driving out the housing oil seal and spacing washer at the same time. Withdraw the shaft from inside the camshaft cover.

27 Examination and renovation: primary drive gears

1 Check the two primary drive gears for worn or broken teeth. If damage of this type is evident, both items should be renewed as a set, since the wear characteristics of one worn gear will soon damage a new component with which it is meshed. The primary driven gear is supplied only complete with the clutch outer drum to which it is riveted. If the rivets work loose they may be tightened using a ball-pein hammer.

2 On original assembly the gears are fitted selectively to ensure that the back lash in the teeth is correct. When renewing gears, ensure that the correctly sized component is fitted. Each gear is marked with a letter A – F, the replacement components should bear the same identification mark as the original part. Failure to match the parts correctly will lead to increased gear noise and accelerated wear.

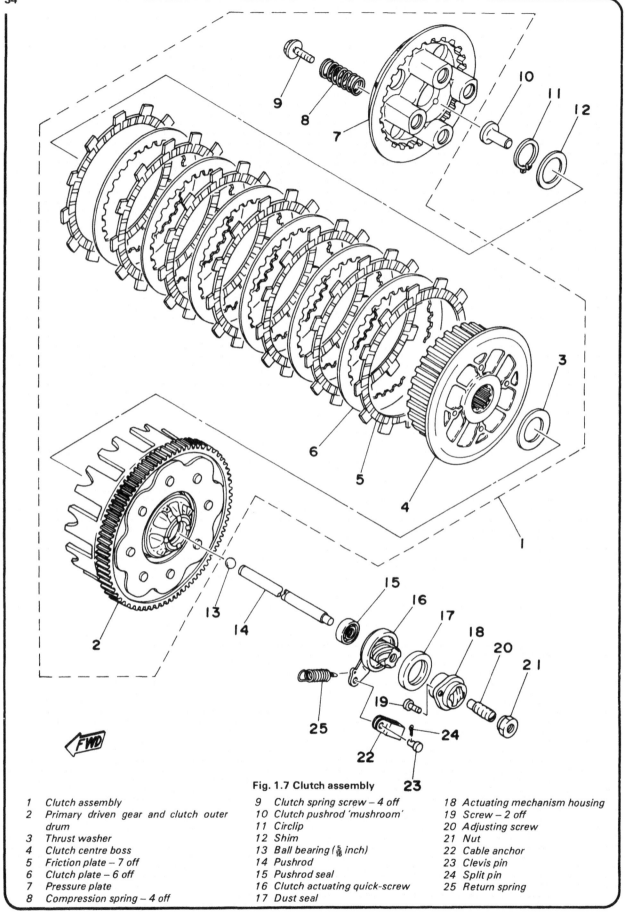

Fig. 1.7 Clutch assembly

1 Clutch assembly
2 Primary driven gear and clutch outer drum
3 Thrust washer
4 Clutch centre boss
5 Friction plate – 7 off
6 Clutch plate – 6 off
7 Pressure plate
8 Compression spring – 4 off
9 Clutch spring screw – 4 off
10 Clutch pushrod 'mushroom'
11 Circlip
12 Shim
13 Ball bearing ($\frac{5}{16}$ inch)
14 Pushrod
15 Pushrod seal
16 Clutch actuating quick-screw
17 Dust seal
18 Actuating mechanism housing
19 Screw – 2 off
20 Adjusting screw
21 Nut
22 Cable anchor
23 Clevis pin
24 Split pin
25 Return spring

26.2a Left-hand rocker pins are held by rubber plugs

26.2b Use casing bolt to withdraw each pin

26.3 Displace 'E' clip to allow removal of shaft

28 Examination and renovation: clutch assembly

1   After an extended period of service the clutch linings will wear and promote clutch slip. The limit of wear measured across each inserted plate and the standard measurement is as follows:

Clutch plate thickness

Standard	Service limit
3 mm (0·12 in)	2·7 mm (0·11 in)

When the overall width reaches the limit, the inserted plates must be renewed, preferably as a complete set.
2   The plain plates should not show any excess heating (blueing). Check the warpage of each plate using plate glass or surface plate and a feeler gauge. The maximum allowable warpage is 0·05 mm (0·002 in).
3   Check the condition of the steel spring rings which lie between the clutch plates. Replace any that are bent or broken.
4   Check the free length of each clutch spring with a vernier gauge. After considerable use the springs will take a permanent set thereby reducing the pressure applied to the clutch plates. The correct measurements are as follows:

Clutch springs

Standard	Service limit
34·6 mm (1·362 in)	33·6 mm (1·323 in)

5   Examine the clutch assembly for burrs or indentation on the edges of the protruding tongues of the inserted plates and/or slots worn in the edges of the outer drum with which they engage. Similar wear can occur between the inner tongues of the plain clutch plates and the slots in the clutch inner drum. Wear of this nature will cause clutch drag and slow disengagement during gear changes, since the plates will become trapped and will not free fully when the clutch is withdrawn. A small amount of wear can be corrected by dressing with a fine file; more extensive wear will necessitate renewal of the worn parts.
6   The clutch release mechanism attached to the final drive sprocket cover does not normally require attention provided it is greased at regular intervals. It is held to the cover by two crosshead screws and operates on the worm and quick start thread principle.

29 Examination and renovation: gearbox components

1   Examine each of the gear pinions to ensure that there are no chipped or broken teeth and that the dogs on the end of the pinions are not rounded. Gear pinions with any of these defects must be renewed; there is no satisfactory method of reclaiming them. Dismantling the two gearshaft assemblies is quite straightforward, requiring the removal of the thrust washers and locating washers to free each pinion in sequence. If necessary, make sketches to facilitate correct reassembly.
2   The gearbox bearings must be free from play and show no signs of roughness when they are rotated. After thorough washing in petrol the bearings should be examined for roughness and play. Also check for pitting on the roller tracks.
3   It is advisable to renew the gearbox oil seals irrespective of their condition. Should a re-used oil seal fail at a late date, a considerable amount of work is involved to gain access to renew it.
4   Check the gear selector rods for straightness by rolling them on a sheet of plate glass. A bent rod will cause difficulty in

selecting gears and will make the gear change particularly heavy.

5   The selector forks should be examined closely, to ensure that they are not bent or badly worn. The case hardened pins which engage with the cam channels are easily renewable if they are worn. Under normal conditions, the gear selector mechanism is unlikely to wear quickly, unless the gearbox oil level has been allowed to become low.

6   The tracks in the selector drum, with which the selector forks engage, should not show any undue signs of wear unless neglect has led to under-lubrication of the gearbox. Check the tension of the gearchange pawl, gearchange arm and drum stopper arm springs. Weakness in the springs will lead to imprecise gear selection. Check the condition of the gear stopper arm roller and the pins in the change drum end with which it engages. It is unlikely that wear will take place here except after considerable mileage.

30 Examination and replacement: kickstart shaft assembly

1   Check the condition of the kickstart components. If slipping has been encountered, a worn ratchet and pawl may be traced as the cause.

2   The same symptom may be a result of the clip being either too tight or too loose on the pinion boss. The condition of the clip may be checked by attaching a spring balance to the clip projection and reading off the force needed to promote slip. A reading of 0·8 – 1·3 kg (1·8 – 2·9 lbs) indicates that the clip is in good condition.

3   Any other damage or wear to the components will be self-evident. If either the ratchet or pawl is found to be faulty, both components must be renewed as a pair. Examine the kickstart return spring, which should be renewed if there is any doubt about its condition.

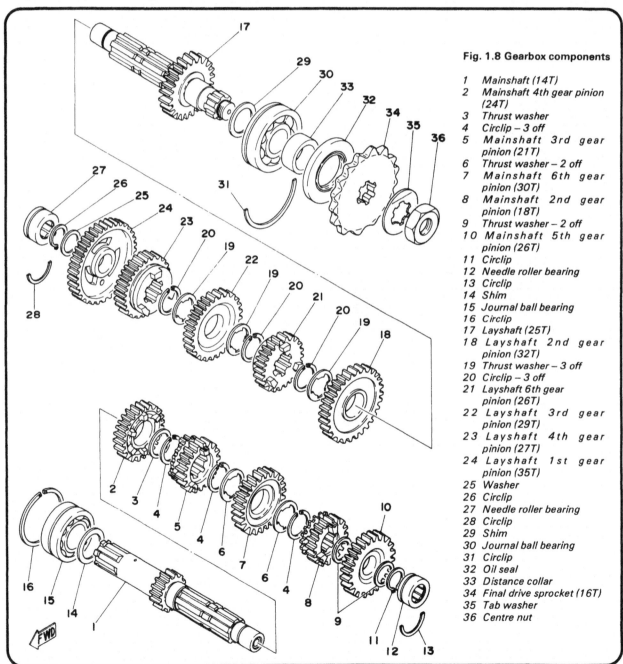

Fig. 1.8 Gearbox components

1   Mainshaft (14T)
2   Mainshaft 4th gear pinion (24T)
3   Thrust washer
4   Circlip – 3 off
5   Mainshaft 3rd gear pinion (21T)
6   Thrust washer – 2 off
7   Mainshaft 6th gear pinion (30T)
8   Mainshaft 2nd gear pinion (18T)
9   Thrust washer – 2 off
10 Mainshaft 5th gear pinion (26T)
11 Circlip
12 Needle roller bearing
13 Circlip
14 Shim
15 Journal ball bearing
16 Circlip
17 Layshaft (25T)
18 Layshaft 2nd gear pinion (32T)
19 Thrust washer – 3 off
20 Circlip – 3 off
21 Layshaft 6th gear pinion (26T)
22 Layshaft 3rd gear pinion (29T)
23 Layshaft 4th gear pinion (27T)
24 Layshaft 1st gear pinion (35T)
25 Washer
26 Circlip
27 Needle roller bearing
28 Circlip
29 Shim
30 Journal ball bearing
31 Circlip
32 Oil seal
33 Distance collar
34 Final drive sprocket (16T)
35 Tab washer
36 Centre nut

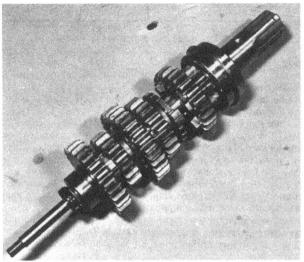

29.1a Assembled mainshaft – general view

29.1b Assembled layshaft – general view

30.1a Check the condition of the kickstart clip

30.1b Insert spring inner turned end into shaft hole and ...

30.1c ... slide in guide to engage turned end

30.1d Secure by means of the circlip

31 Engine/gearbox reassembly: general

1   Before reassembly of the engine/gear unit is commenced, the various component parts should be cleaned thoroughly and placed on a sheet of clean paper, close to the working area.
2   Make sure all traces of old gaskets have been removed and that the mating surfaces are clean and undamaged. One of the best ways to remove old gasket cement is to apply a rag soaked in methylated spirit. This acts as a solvent and will ensure that the cement is removed without resort to scraping and the consequent risk of damage.
3   Gather together all the necessary tools and have available an oil can filled with clean engine oil. Make sure all new gaskets and oil seals are to hand, also all the replacement parts required. Nothing is more frustrating than having to stop in the middle of a reassembly sequence because a vital gasket or replacement has been overlooked.
4   Make sure that the reassembly area is clean and that there is adequate working space. Refer to the torque and clearance settings wherever they are given. Many of the smaller bolts are easily sheared if over-tightened. Always use the correct size screwdriver bit for the crosshead screws and never an ordinary screwdriver or punch. If the existing screws show evidence of maltreatment in the past, it is advisable to renew them as a complete set.

32 Engine/gearbox reassembly: replacing the gear selector mechanism

1   Insert the gearchange drum through the aperture in the gearbox wall so that the drum projects into the gearbox. Replace the drive peg in the end of the change drum and then fit the drum stopper plate so that it locates with the peg. Secure the plate by means of the circlip. The circlip must be fitted so that the two eye ends are positioned as shown in the accompanying photograph. In any other position the circlip will foul the detent plunger. Lubricate the drum bearing surfaces and

then slide the unit into place in the gearbox left-hand wall.
2   Invert the crankcase lower half and refit the detent plunger, spring and housing bolt so that the plunger locates with the change drum. Check the bolt sealing washer before refitting. Rotate the change drum so that it is in the neutral position. This can be found by turning the change drum fully anti-clockwise and then one position back.
3   Place the crankcase lower half upright on the workbench. Insert the longer of the two selector fork rods into the gearbox wall forward of the change drum. Position the two selector forks and push the rod fully home. The forks should be placed as shown in the accompanying photograph, with the left-hand one locating with the left-hand cam track and the right-hand one locating with the extreme right-hand cam track. Fit the rear selector forks in a similar manner, noting that the left-hand fork should locate with centre cam track. Fit the retaining 'E' clip to the rear selector fork rod groove adjacent to the gearbox left-hand wall and then fit the blind grommets to the rod ends.

33 Engine/gearbox reassembly: replacing the gearshaft assemblies

1   Before being fitted into the crankcase, the two gearshafts must be reassembled complete with the various pinions, washers, circlips and bearings. Assemble each shaft by referring to the relevant illustrations, ensuring that all components are fitted in the correct sequence. Even one misplaced washer or circlip may lead to incorrect gear selection and the need for considerable dismantling at a later stage.
2   Fit the two small and one large bearing locating half clips into the bearing recesses in the casing. Lower the mainshaft into position, followed by the layshaft, which should have the left-hand oil seal and collar in place on the shaft. Ensure that the bearings enter the locating clips and that the selector forks engage with the channels in the gear pinions.
3   Place the pushrod oil seal in the recess to the left of the mainshaft end.

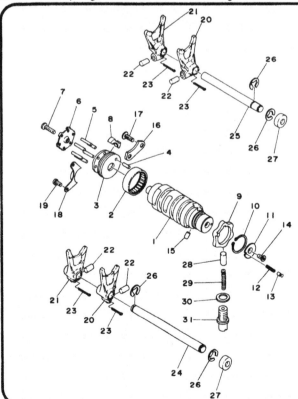

Fig. 1.9 Gear selector mechanism

1	Gearchange drum	17	Countersunk screw – 2 off
2	Needle roller bearing	18	Change lever guide
3	Pin carrier	19	Screw – 2 off
4	Drive pin	20	Selector fork – 2 off
5	Change pin – 6 off	21	Selector fork – 2 off
6	End plate	22	Drum follower pin – 4 off
7	Countersunk screw	23	Split pin – 4 off
8	Pawl lifter	24	Selector fork rod
9	Stopper plate	25	Selector fork rod
10	Circlip	26	Circlip – 4 off
11	Neutral indicator switch plate	27	Plug – 2 off
12	Spring	28	Detent plunger
13	Contact piece	29	Spring
14	Countersunk screw	30	Sealing washer
15	Drive pin	31	Detent housing
16	Guide plate		

32.1a Insert change drum and locate stopper plate with peg

32.1b Circlip must be positioned, as shown

32.2 Replace the detent housing bolt, plunger and spring

32.3a Fit the guide pins to the selector forks

32.3b Locate the forward selector forks and insert the rod

32.3c Fit the rear selector forks and ...

32.3d ... secure the rod by means of the 'E' clip

32.3e Drift both rod blind grommets into place

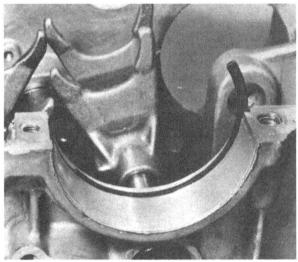

33.2a Slide the gearbox bearing half clips into position

33.2b Fit completed mainshaft and then replace ...

33.2c ... the completed layshaft with left-hand seal in position

33.3 Install the clutch pushrod seal in end of the mainshaft

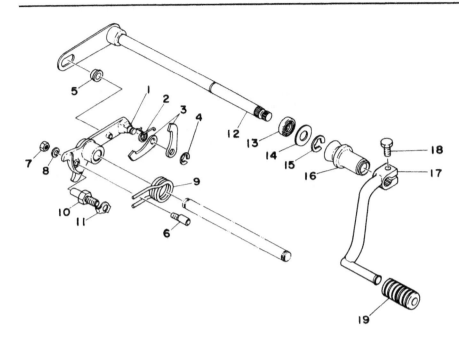

Fig. 1.10 Gearchange mechanism

1   Gear selector arm
2   Pawl spring
3   Gearchange pawl – 2 off
4   Circlip
5   Gear selector arm roller
6   Centraliser screw
7   Nut
8   Spring washer
9   Return spring
10 Stop
11 Lock washer
12 Gearchange shaft
13 Oil seal
14 Washer
15 Circlip
16 Sealing rubber
17 Gearchange lever
18 Pinch bolt
19 Rubber

34 Engine/gearbox reassembly: replacing the crankshaft

1   Replace the oil baffle plates in the crankcase, depressing them firmly to enable the transverse retaining rod to be fitted.
2   Clean the main bearing housings thoroughly and then refit the new main bearing shells into the two crankcase halves. The locating tags must engage with the relieved portion of each housing and the ends of each shell should be flush with the crankcase mating surface.
3   Fit the cam chain onto the crankshaft central sprocket and then lubricate the main bearing journals. Lift the crankshaft up into position in the casing. Check that the main bearing shells have not been displaced.

35 Engine/gearbox reassembly: joining the crankcase halves

1   Ensure that both mating surfaces of the crankcase halves are clean and smear them with a light coating of gasket cement. Check again that the bearing shells are located correctly.

2   Fit the upper crankcase half into position on the lower half so that it locates with the dowel pins and gently tap it down with the flat of the hand. Check that the mating surfaces have contacted all round the cases. If any difficulty is encountered in fitting the two cases together, do not use force. Remove the lower half, check the components for correct positioning and then repeat the fitting operation.
3   Refit the bolts into the crankcase upper half, bearing in mind that the rear bolt behind and to the right of the breather chamber also retains the main earth lead. Do not omit the clips which secure the alternator leads. Tighten the bolts only by hand at this stage. Invert the crankcase and fit the lower bolts. Tighten all the bolts evenly, a little at a time, in the sequence given in the accompanying illustration. The final torque settings are as follows:

8 mm bolts	2·0 – 2·4 kg m (lb ft)
6 mm bolts	0·8 – 1·2 kg m (lb ft)

During tightening, check that the crankshaft and engine shafts revolve freely.

34.3 Lift crankshaft and camchain into position in crankcase

35.2 Lower upper casing into place over gearshafts and crankshaft

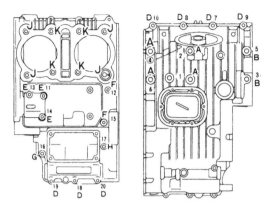

Fig. 1.11 Crankcase bolt tightening sequence

36 Engine/gearbox reassembly: replacing the gearchange mechanism

1 Fit the change drum guide plate and selector arm guide plate into the primary chaincase, tightening the screws fully. Fit the centraliser spring onto the selector arm boss as shown in the accompanying photograph. Fit the selector arm into the casing, simultaneously holding apart the two change pawls so that they clear the end of the change drum. When released, the pawls should locate with the change pins. Secure the selector arm by means of the 'E' clip.

2 Place the shouldered pivot bush onto the stub projecting from the selector arm so that the smaller diameter is outwards. Lubricate the splined end of the gearchange shaft and then insert it into position in the casing, pushing it fully home so that the elongated hole in the shaft arm engages with the shouldered pivot bolt. From the opposite side of the engine, fit the thrust washer and 'E' clip which secure the gearchange shaft.

3 Before continuing with assembly, temporarily replace the gear change lever and attempt to select each gear in turn. Rotating the two gearshafts will aid selection. If difficulty is encountered in selection and no obvious cause is evident, the crankcase halves must be separated and the fault traced. Incorrect assembly of gear clusters or selector forks should be looked for. Sluggish gear selection may be caused by incorrect centralisation of the gear selector arm, which should be checked in any case as a matter of course.

4 Place the engine in first gear and check that the distance between each of the two selector pawls and its adjacent change pin is equal. See the accompanying illustration. If the distances are unequal, slacken the locknut on the adjuster screw in the selector arm and rotate the screw as necessary, to equalise the pawls. Tighten the locknut without allowing the screw to turn. Carry out this check in each gear, if necessary making further adjustment. It may be necessary to make a compromise in adjustment if any components are worn.

37 Engine/gearbox reassembly: replacing the primary drive gear, clutch and kickstart shaft

1 Fit the Woodruff key into the right-hand end of the crankshaft so that the top face is parallel with the shaft. Slide the primary drive pinion onto the shaft so that the internal keyway engages with the key. The pinion must be fitted with the raised boss facing outwards. Replace the centre bolt and washer with the convex face of the washer away from the pinion. Use a close fitting bar through one small-end eye to lock the crankshaft and tighten the bolt to $4 \cdot 0 - 4 \cdot 5$ kg m (29 – 33 lb ft).

2 Place the clutch outer drum over the clutch shaft (mainshaft) after lubricating the bearing surface. Grease the clutch pushrod and insert it into the hollow shaft with the shouldered end towards the left-hand side. Similarly fit the steel ball and the clutch operating thrust piece.

3 Install the clutch centre backing washer over the clutch shaft and then fit the centre boss, securing it by means of the large washer and circlip. Check that the circlip is located securely in the groove. Replace the clutch plates, commencing with a friction plate followed by a plain plate. Continue fitting the plates alternately and then replace the clutch pressure plate, springs and bolts. Note that the pressure plate is marked with a single arrow at the periphery. This arrow should be aligned with a similar mark on the outer raised face of the centre boss. Tighten the clutch bolts fully.

4 If the kickstart shaft assembly was dismantled for inspection or attention it should now be reassembled, prior to installation in the casing. Refer to the appropriate line drawing for the correct assembly sequence. Insert the completed unit into the casing, ensuring that the clip projection enters the guideway in the lower edge of the casing. Grip the outer turned end of the kickstart return spring with a stout pair of pliers and tension the spring in a clockwise direction through approximately one full turn, until the hook end can be anchored on the pillar projecting from the casing wall.

5 Temporarily refit the kickstart lever and check that the starting mechanism functions correctly.

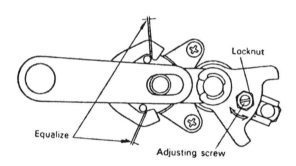

Fig. 1.12 Centralising the gearchange pawls

36.1a Install the drum guide plate and selector arm guide

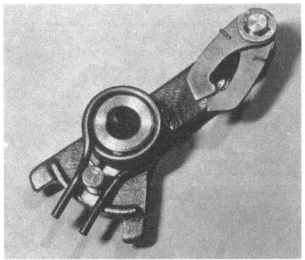

36.1b Main centraliser spring must be fitted as shown

36.1c Replace the 'E' clip to retain the change arm

36.2a Place the pivot bush on the change arm stub and ...

36.2b ... slide the gearchange shaft into place locating with the bush

36.2c Secure the gearchange shaft with the washer and 'E' clip

37.1 Install the primary drive pinion with the boss outermost

37.2a Fit the clutch drum and thrust washer onto the shaft

37.2b Insert the clutch pushrod and ...

37.2c ... the single steel ball, followed by ...

37.2d ... the clutch thrust piece

37.3a Replace the clutch centre boss and ...

37.3b ... the heavy plate washer

37.3c Ensure that the circlip is correctly seated in the groove

37.3d Install the clutch plates alternately and ...

37.3e ... fit the clutch pressure plate

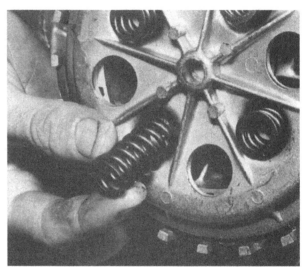

37.3f Replace the clutch springs and ...

37.3g ... the bolts which should be tightened fully

37.4 Tension kickstart spring in clockwise direction

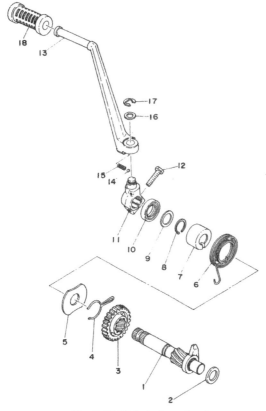

Fig. 1.13 Kickstart mechanism

1 Kickstart shaft	10 Oil seal
2 Shim	11 Splined boss
3 Kickstart pinion (23T)	12 Pinch bolt
4 Pinion clip	13 Kickstart crank
5 Spring cover	14 Ball ($\frac{7}{32}$ inch)
6 Return spring	15 Compression spring
7 Spring guide	16 Washer
8 Circlip	17 'E' clip
9 Washer	18 Pedal rubber

38 Engine/gearbox reassembly: replacing the oil pump, primary drive cover and the oil strainer screen

1   The oil pump should be refitted as a completed unit, together with the oil pump idler gear. Check that the single dowel is in position in the rear part of the casing and place the oil pump in position. Insert the three socket bolts, tightening them until it is just possible to move the pump up and down slightly in a vertical direction. The oil pump is fitted with a single dowel only, so that its position may be altered and the backlash between the pump idler pinion and the primary drive pinion can be adjusted. To make the adjustment, select a 0·4 mm (0·015 in) feeler gauge and slide it between the driven sides of the central meshed teeth of the two pinions. The position of the feeler gauge will be just to the right of an imaginary line drawn between the centres of the two pinions. Push the oil pump forward bracket arm upwards so that the feeler gauge is gripped lightly and then tighten the forward socket bolt. Tighten the remaining two bolts and recheck the backlash.
2   Lubricate liberally with engine oil the components within the primary drive chamber and fit a new gasket over the locating dowels. Grease the lip of the kickstart shaft oil seal and check that the kickstart shaft thrust washer is in place. Fit the primary drive cover and tighten the screws evenly, in a diagonal sequence.
3   Tilt the engine forwards so that access can be made to the

crankcase base. Install the oil strainer screen and fit the cover, together with a new gasket. The cover should be fitted with the arrow mark pointing forwards.

39 Engine/gearbox reassembly: replacing the alternator and stater motor

1   Lubricate the starter clutch driven sprocket bush and slide the sprocket onto the left-hand end of the crankshaft. Fit the sprocket and replace the retention tongue which is held by a single bolt. The tongue should be fitted with the turned upper end abutting against the flat milled on the casing boss.
2   Replace the small Woodruff key in the tapered shaft end and install the alternator rotor so that the internal keyway engages with the key. In order to allow the starter driven sprocket boss to enter the starter clutch rollers, rotate the sprocket clockwise as the rotor is being pushed home. Fit the alternator bolt, tightening it to 3·0 – 3·5 kg m (22 – 25 lb ft).
3   Loop the starter chain around the sprocket and mesh the starter motor sprocket towards the rear of the casing. The sprocket should be fitted with the tapered boss outwards. If the starter motor lead was disconnected from the motor, it must be refitted now, before installation of the starter motor itself. Lightly grease the starter motor boss 'O' ring and insert the unit into the casing. Locate the sprocket on the splined motor shaft and push the motor fully home. Insert and tighten the two retaining bolts. Push the wiring lead and grommet into the rebate in the side of the casing and fit the starter motor cover
4   Place a new gasket on the alternator cover locating dowels and fit the cover, complete with the alternator stator. Fit and tighten the retaining screws.

40 Engine/gearbox reassembly: replacing the neutral indicator switch and oil pressure warning switch

1   Check the condition of the 'O' ring on the neutral indicator switch and then fit the switch into the casing. The switch should be positioned with the terminal screw towards the rear of the engine.
2   Apply a small quantity of sealing compound to the tapered threads of the oil pressure switch. Insert and tighten the switch. Do not overtighten or the switch may shear off. Reconnect the wires to the two switches, by means of the small screws.
3   Track the two alternator leads down behind the neutral indicator switch and through the recess in the bottom edge of the casing. Secure the cables by means of the two cable clamps.

38.1 Install the oil pump complete with idler gear and dowel

38.2 Use a new gasket at the primary drive cover

38.3 Refit the oil strainer, gasket and cover

39.1a Install the sprocket after lubricating the bush

39.1b Secure the sprocket by means of the guide

39.2a Fit the Woodruff key in the crankshaft and ...

39.2b ... slide the alternator rotor into postion

39.3a Fit the starter sprocket with the tapered boss outwards

39.3b Install the starter motor together with the cable

39.3c Fit the alternator cover, using a new gasket

40.1 Do not omit the 'O' ring on the neutral switch

40.2 DO NOT overtighten the oil pressure switch

41 Engine/gearbox reassembly: refitting the pistons and cylinder block

1   Place the engine in the normal upright position so that it rests securely on the workbench.

2   Place clean rags in the crankcase mouths, around the connecting rods to prevent any small components, such as circlips, from falling down into the crankcase.

3   Carefully fit the rings on to each piston. The oil ring should be fitted from the skirt of the piston and the second compression ring fitted over the crown. Make certain that the compression rings are fitted the correct way up with the stamped side facing upwards.

4   Fit each piston on to its respective small end after thoroughly lubricating the gudgeon pin with clean engine oil. Make certain that the correct piston is fitted to the respective connecting rod and that it is the proper way round, ie; with the arrow on the crown pointing forwards. If the gudgeon pin is a tight fit in the piston bosses, pre-heat the piston in boiling water; do not force the gudgeon pin into place. Ensure that the piston is absolutely dry before fitting. Replace the circlips making sure that they seat correctly in their retaining grooves.

Warning: Never re-use old circlips since they are likely to be stretched or weakened. A loose circlip will cause extensive damage to the cylinder bore and piston. Arrange the ring gaps so that they are at about 120° to each other.

5   Turn the crankshaft until both pistons are at equal height. Place a new cylinder base gasket over the holding down studs. Gasket cement should not be necessary. Fit the two forward locating dowels and the large dowel on the rear right-hand cylinder holding down stud around which is fitted an 'O' ring. Check that the large 'O' rings are correctly positioned on the cylinder sleeves, where they leave the cylinder block, and liberally lubricate the cylinder bores with engine oil. Hook the camshaft chain up and secure it with a screwdriver or wire. Check it is still engaged with the crankshaft sprocket.

6   Position the cylinder block over the holding down studs and slide it down until the pistons begin to enter the cylinder bores. Piston ring clamps should not be required as the cylinder sleeves have a generous lead in, but care should be taken that the ring ends do not become dislodged from the ring grooves. Hook the camshaft chain up through the tunnel between the cylinder bores and place a screwdriver through the chain, at the top. Remove the rag padding from the crankcase mouths and slide the cylinder block fully home.

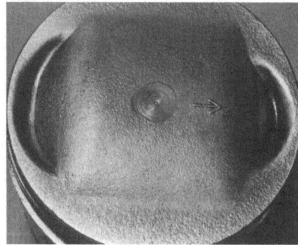

41.4a Piston must be fitted with arrow pointing forwards

41.4b Insert gudgeon pin and fit ...

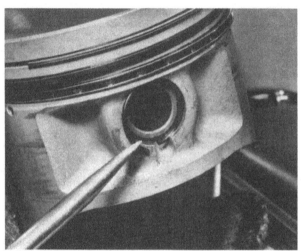

41.4c ... the circlips, ensuring that they seat correctly

41.5a Fit the large dowel and the seal and ...

41.5b ... install a new base gasket and the two dowels

41.5c Check that cylinder 'O' rings are correctly positioned

41.6 Draw chain up through cylinder block

42 Engine/gearbox reassembly: replacing the cylinder head

1   Before the cylinder head is refitted, the valves and associated components must be replaced. Reassemble the valve and valve springs by reversing the dismantling procedure. Fit new oil seals to each valve guide and oil both the valve stem and the valve guide, prior to reassembly. Take special care to ensure the valve guide oil seal is not damaged when the valve is inserted. As final check after assembly, give the end of each valve stem a light tap with a hammer, to make sure the split collets have located correctly. The valve springs are more closely wound at one end than the other. The springs must be fitted with the close wound pitch towards the cylinder head.

2   Fit a new cylinder head gasket to the top of the cylinder block; no gasket cement is required. Replace the two front dowels and the right-hand rear dowel and rubber seal.

3   Loop a length of wire around the camshaft chain and remove the screwdriver or rod. Position the cylinder head so that the wire can be run through the chain tunnel and slide the head down over the holding down studs. Place a screwdriver through the chain.

4   Replace the thick washers over the holding down studs and fit the dome nuts and also the two flange bolts either side of the plug holes. The cylinder head holding down nuts and bolts must be tightened down very evenly, otherwise there is considerable danger of distorting the large aluminium casting. Tighten the head nuts and the two bolts in the sequence given in the accompanying illustration. The torque figures are as follows:

    10 mm bolts   3·0 – 3·5 kg m (21·7 – 24·6 lb ft)
    6mm bolts   0·8 – 1·2 kg m (5·2 – 8·0 lb ft)

43 Engine/gearbox reassembly: replacing the camshaft and tensioner and timing the valves

1   Apply a spanner to the alternator rotor centre bolt and turn the engine in a forward direction until the LT mark on the rotor aligns with the index mark on the alternator cover. The cam chain must be hand fed to prevent it bunching or jamming on the lower sprocket. The left-hand piston is now at TDC, the datum position from which valve timing is made.

2   Position the camshaft sprocket to the right of the cam chain so that the two marked lines on the sprocket face are to the left-hand side. Insert the camshaft from the left, through the sprocket and chain. The sprocket has a slightly raised face around both bolt holes, of which one has a small raised pip on the outer edge. Mesh the sprocket with the chain so that the

bolt hole pip points vertically from the engine and so that the two lines marked on the sprocket face are parallel with the mating surface of the central bearing housing. Push the camshaft fully into position and then rotate it so that the ATU drive pin in the camshaft end is at the top. Insert the first sprocket retaining bolt and tighten it slightly. Before proceeding further, check that the LT mark on the alternator rotor is still in line with the mark and that the camshaft and sprocket are still in the prescribed positions. If all is well, rotate the crankshaft to gain access to the second sprocket bolt hole and insert the bolt. Both sprocket retaining bolts should be secured by applying a small amount of locking fluid to the threads. Tighten the bolts evenly, turning the engine to gain access as required.

3   Insert the cam chain guide blade into the front of the chain tunnel, ensuring that it locates at the lower end with the recess in the crankcase. Fit the cam chain tensioner blade in a similar manner. The chain tensioner unit is spring loaded, with a ratchet and pawl arrangement to prevent the plunger from being pushed back up into the body after it has moved down to tension the chain automatically. Before fitting the tensioner, depress the pawl and push the plunger in as far as possible. The plunger should be temporarily secured in this position by screwing a suitable small bolt or screw into the tiny threaded hole provided in the upper end of the tensioner body. Fit the tensioner so that the blade locates with the recessed plunger end and tighten the two bolts. The screw or bolt can now be removed to allow the plunger to move downwards and so tension the chain..

4   Fit and tighten the two spark plugs.

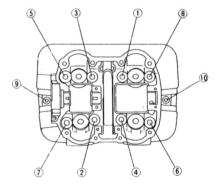

Fig. 1.14 Cylinder head nut & bolt tightening sequence

42.1a Fit the spring seat before fitting the new oil seal

42.1b Lubricate the valve stem and guide thoroughly

42.1c Install the spring with close wound pitch downward

42.1d Compress the springs to allow collet replacement

42.2 Fit a new head gasket, the dowels and dowel seal

42.3 Hook the cam chain up whilst lowering the cylinder head

42.4 Do not omit to tighten the two outer head bolts

43.2a Insert the camshaft to locate the chain and sprocket

43.2b Position camshaft with drive pin at 12 o'clock

43.2c Mesh sprocket with chain so that bolt hole 'pip' is pointing upwards

43.2 Insert and tighten sprocket retaining bolts

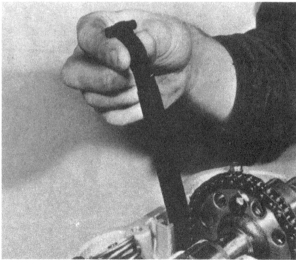

43.3a Refit the chain guide and ...

43.3b ... the tensioner blade

43.3c Compress the tension plunger, securing it with screw

43.3d Replace the chain tensioner before removing the screw

and does not distort the cover. The correct torque setting is as follows:

8mm bolts   2·0 – 2·4 kg m (14·5 – 17·0 lb ft)
6 mm bolts   0·8 – 1·2 kg m (6·0 – 8·5 lb ft)

45 Engine/gearbox reassembly: replacing the automatic timing unit and contact breaker assembly

1   Fit the automatic timing unit on to the end of the camshaft, locating it with the dowel pin. Insert and tighten the ATU retaining bolt.
2   Position the contact breaker base plate, complete with the two contact breaker assemblies, into position over the ATU. Fit the wiring lead grommet into the recess in the edge of the casing and replace the base plate clamping screws finger tight. If, as suggested during dismantling, a line was scribed on the base plate to align with a mark on the casing, the ignition timing will only require checking as described in Chapter 3, Section 5, paragraphs 1 – 2. If the timing is to be carried out from scratch, refer to Chapter 3, Section 5, paragraphs 3 – 5.
3   For the purpose of ignition timing the contact breaker cover and the alternator end cover should be left off until after the engine has been installed in the frame.

44 Engine/gearbox reassembly: replacing the camshaft cover and tachometer driveshaft

1   If the rocker arms and pins have been removed they must now be replaced, together with the tachometer driveshaft, if this too has been displaced. Lubricate the pins before inserting them, with the threaded ends outwards to enable easy removal at any subsequent overhaul. Check the condition of the sealing washers on the plugs holding the right-hand pins.
2   Insert the tachometer driveshaft and housing followed by the plain washers. Carefully drive in the shaft oil seal and then fit the 'E' clip to the housing inside the camshaft cover.
3   Apply a non-hardening sealing compound to the mating surface of the camshaft cover. Check that the two locating dowels are in position in the cylinder head. Lubricate the camshaft and valve thoroughly and partially fill the cam feed troughs with engine oil. Place the camshaft cover in position and insert the bolt. Note that the two bolts fitted on the extreme left-hand side of the camshaft cover each have a single aluminium washer.
4   Tighten the bolts very evenly, in a diagonal sequence, so that the uneven upward pressure from the camshaft is spread

44.3a Use gasket cement at camshaft cover joint

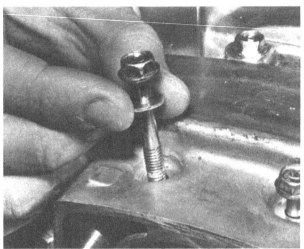

44.3b Note aluminium washers on two left-hand bolts

46.1 Check valve clearances using a feeler gauge

46 Engine/gearbox reassembly: adjusting the valve clearance

Note: *The following procedure assumes that the engine is out of the frame and thus drained of oil. If not, proceed as described in Routine maintenance.*

1   Remove the four inspection caps from the camshaft cover. Note that the engine will be easier to turn over if the spark plugs are removed. Rotate the crankshaft anti-clockwise until the left-hand cylinder inlet valve has opened and closed (sunk down and risen again), then rotate the crankshaft further until the alternator LT mark aligns with the crankcase pointer. The left-hand cylinder is then at TDC on the compression stroke (both valves closed), with free play at both rockers.
2   Measure the clearance by inserting a feeler gauge between the rocker arm adjuster screw and the tip of the valve stem; if the gap is correct a gauge of the specified thickness (see Specifications) will be a tight slip fit. To adjust the clearance, slacken the adjuster locknut and screw the adjuster in or out as necessary, then hold the screw steady while the locknut is tightened. Tighten the locknut securely but be careful not to overtighten it as this merely distorts the thread and makes future adjustment very difficult. Recheck the clearance to ensure that the setting has not altered, then repeat the procedure on the remaining valve.
3   When the left-hand cylinder valve clearances are correct turn the crankshaft forwards until the right-hand cylinder is positioned, as described above, at TDC on the compression stroke (alternator rotor RT mark aligned and free play at both rockers) then check the clearances of the right-hand valves.
4   Check that the spark plugs are clean and correctly gapped and refit them. Refit the four valve inspection caps but do not fit the alternator inspection cover until the ignition timing has been checked.

47 Replacing the engine/gearbox unit in the frame

1   As is the case with removal, engine replacement requires considerable care and patience. Replacing the engine necessitates the use of two people and it is important that the machine is standing firmly on level ground. Lift the engine in from the right-hand side of the machine, with the front of the engine going in first. Tip the engine over to the left and then lift the rear of the engine up and into place.
2   Lift the engine when necessary to stop the bolt threads fouling the brackets and insert the bottom rear engine bolt from the left. Fit the two front bolts, again lifting the engine, if required. Replace the two engine mounting brackets to the rear

of the engine, remembering that the stop lamp switch bracket is held by the right-hand bracket lower bolt. Before replacing the nuts on the engine mounting bolts, refit the head steady bracket which is held to the camshaft cover. The head steady is held to the cover by a single through bolt and to the frame top tube by two bolts. Fit and tighten all the mounting bolt nuts, commencing with the lowest and finishing at the head steady.
3   Track the starter motor lead across behind the engine, from left-to-right, and secure it to the mudguard before reconnection to the starter solenoid upper terminal. Route the two cables from the alternator up forward of the battery box, securing them by means of the clips provided, and reconnecting them at the two block connectors. Reconnect the two contact breaker leads.
4   Rest the air hose unit on the rear of the crankcase, close to the frame down tubes. Position the carburettors to the right-hand side of the machine and reconnect the throttle cable to the pulley wheel between the carburettors. Slide the carburettors into place and then forwards, so they enter the inlet stubs. Ensure that the carburettors are correctly positioned before tightening the screw clips. Manoeuvre the air hose unit upwards so that it may be pushed into place on the carburettor mouths. Replace the air filter boxes and retaining straps and then tighten the four screw clips holding the hose unit. Reconnect the breather pipe at the breather cover.
5   Replace the exhaust system after fitting a new gasket ring into each exhaust port. Tighten the flange nuts evenly and then tighten the silencer/pipe joint clamps.
6   Fit the final drive sprocket to the gearbox splined shaft. Replace the tab washer and the nut, which should be placed with the relieved face towards the sprocket. Loop the final drive chain over the engine sprocket and mesh the two ends onto the rear wheel sprocket. Replace the master link making certain that the spring link is replaced the correct way round. That is with the closed end facing the direction of travel. Apply the rear brake to prevent rotation of the sprocket and then tighten the centre nut fully. Bend up the tab washer to secure the nut.
7   Reconnect the clutch cable with the lifting mechanism in the final drive sprocket cover. Do not omit the cable abutment piece which fits into the top of the case. Check that the lifting mechanism is well lubricated with grease and that the small central steel ball is in place. Fit the gearchange shaft protecting boot and then replace the cover and screws. Refit the gearchange pedal, checking the operating angle before tightening the bolt.
8   Replace both front footrests. A small amount of household detergent applied to the rubber mounting dampers will aid replacement.
9   Reconnect the tachometer drive cable at the cylinder head, securing it with the single cross-head screw.

10 Replace the battery and reconnect it to the electrical system. Ensure that the red lead is connected to the positive (+) terminal and the black lead is connected to the negative(−) terminal. Check also that the breather tube is reconnected and routed so that the tube end is clear of any cycle parts. Vented electrolyte will quickly corrode components with which it comes in contact.

11 Refit the petrol tank and connect the fuel lines. Do not omit the small spring clips.

12 Reconnect the spark plug caps to their respective spark plugs. Give a final visual check to all electrical connections and replace the two side covers. Both are a push fit.

13 Fit a new oil filter element into the filter chamber, after first inserting the oil filter spring and spacer washer. Check the condition of the sealing 'O' ring and then position the chamber against the front of the crankcase. Tighten the oil filter bolt to 1·3 – 1·7 kg m (9·5 – 12·0 lb ft). Replenish the engine with SAE 20W/50 engine oil. It will require 2·6 litres (5·5/4·6 US/Imp pts). Allow the oil level to settle for a few moments and then check the level by means of the dipstick integral with the filler cap. Do not screw the cap in when checking the level; allow the cap to rest on the casing edge. Kick the engine over smartly with the ignition off, to help prime the oilways.

47.3 Alternator connectors cannot be incorrectly reconnected

47.5a Fit a new ring gasket to each exhaust port

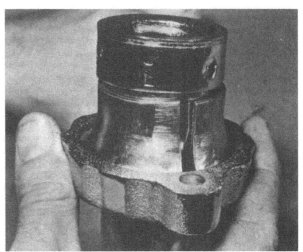

47.5b Split collars should be fitted as shown

47.6a Replace the sprocket nut with relieved face inwards

47.6b Bend up tab washer to secure the nut

47.7a Check that the steel ball is fitted to the clutch

47.7b ... reconnect the clutch cable

48 Starting and running the rebuilt engine

1   Open the petrol tap, close the carburettor chokes and start the engine, using either the kickstart or the electric starter. Raise the chokes as soon as the engine will run evenly and keep it running at a low speed for a few minutes to allow oil pressure to build up and the oil to circulate. If the red oil pressure indicator lamp is not extinguished, stop the engine immediately and investigate the lack of oil pressure.
2   The engine may tend to smoke through the exhausts initially, due to the amount of oil used when assembling the various components. The excess of oil should gradually burn away as the engine settles down.
3   Check the exterior of the machine for oil leaks or blowing gaskets. Make sure that each gear engages correctly and that all the controls function effectively, particularly the brakes. This is an essential last check before taking the machine on the road.

49 Taking the rebuilt machine on the road

1   Any rebuilt machine will need time to settle down, even if

parts have been replaced in their original order. For this reason it is highly advisable to treat the machine gently for the first few miles to ensure oil has circulated throughout the lubrication system and that any new parts fitted have begun to bed down.
2   Even greater care is necessary if the engine has been rebored or if a new crankshaft has been fitted. In the case of a rebore, the engine will have to be run-in again, as if the machine were new. This means greater use of the gearbox and a restraining hand on the throttle until at least 500 miles have been covered. There is no point in keeping to any set speed limit; the main requirement is to keep a light loading on the engine and to gradually work up performance until the 500 mile mark is reached. These recommendations can be lessened to an extent when only a new crankshaft is fitted. Experience is the best guide since it is easy to tell when an engine is running freely.
3   If at any time a lubrication failure is suspected, stop the engine immediately, and investigate the cause. If an engine is run without oil, even for a short period, irreparable engine damage is inevitable.
4   When the engine has cooled down completely after the initial run, recheck the various settings, especially the valve clearances. During the run most of the engine components will have settled into their normal working locations.

50 Fault diagnosis: engine

Symptom	Cause	Remedy
Engine will not start	Defective spark plugs	Remove the plugs and lay on cylinder head. Check whether a good spark occurs when ignition is switched on and engine rotated.
	Dirty or closed contact breaker points	Check condition of points and whether gap is correct.
	Faulty or disconnected condenser	Check whether points arc when separated. Renew condenser if evidence of arcing, or if a weak spark is obtained and the plugs are in good condition.
Engine runs unevenly	Ignition and/or fuel system fault	Check each system independently, as though engine will not start.
	Blowing cylinder head gasket	Leak should be evident from oil leakage where gas escapes.
	Incorrect ignition timing	Check accuracy and if necessary reset.

Lack of power	Fault in fuel system or incorrect ignition timing	See above.
Heavy oil consumption	Cylinder block in need of rebore	Check for bore wear, rebore and fit oversize pistons if required.
	Damaged oil seals	Check engine for oil leaks.
	Worn valve guides and/or stems	Check for wear and renew if necessary.
Excessive mechanical noise	Worn cylinder block (piston slap)	Rebore and fit oversize pistons.
	Worn camshaft drive chain (rattle)	Renew chain.
	Worn big end bearings (knock)	Fit replacement big-end shells.
	Worn main bearings (rumble)	Fit replacement main-bearing shells.
Engine overheats and fades	Lubrication failure	Stop engine and check whether internal parts are receiving oil. Check oil level in crankcase.

51 Fault diagnosis: clutch

Symptom	Cause	Remedy
Engine speed increased as shown by tachometer but machine does not respond	Clutch slip	Check clutch adjustment for free play at handlebar lever. Check thickness of inserted plates.
Difficulty in engaging gears. Gear changes jerky and machine creeps forward when clutch is withdrawn. Difficulty in selecting neutral	Clutch drag	Check clutch adjustment for too much free play. Check clutch drums for indentations in slots and clutch plates for burrs on tongues. Dress with file if damage not too great.
Clutch operation stiff	Damaged, trapped or frayed control cable	Check cable and renew if necessary. Make sure cable is lubricated and has no sharp bends.
	Bent operating pushrod	Check the pushrod for trueness.

52 Fault diagnosis: gearbox

Symptom	Cause	Remedy
Difficulty in engaging gears	Selector forks bent	Renew.
	Gear clusters not assembled correctly	Check gear cluster arrangement and position of thrust washers.
Machine jumps out of gear	Worn dogs on ends of gear pinions	Renew worn pinions.
	Stopper arms not seating correctly	Remove right hand crankcase cover and check stopper arm action.
Gearchange lever does not return to original position	Broken return spring	Renew spring.
Kickstart does not return when engine is turned over or started	Broken or poorly tensioned return spring	Renew spring or re-tension
Kickstart slips	Ratchet assembly or kickstart clip worn	Renew worn components.

Chapter 2 Fuel System and Lubrication

For information relating to later models, see Chapter 7

Contents

Specifications

Fuel tank capacity ... ... 11.0 litres (2.9/2.4 US/Imp gallons)

Engine oil capacity

Total	2.6 litres (5.5/4.6 US/Imp pints)
At oil change	2.0 litres (4.2/3.5 US/Imp pints)
At oil and filter change	2.3 litres (5.0/4.0 US/Imp pints)

Carburettor

	XS250	XS360	XS400
Type	BS32	BS34	BS34
Make	Mikuni	Mikuni	Mikuni
Main jet	117.5	135	142.5
Air jet	0.6	0.6	45
Jet needle	4Z1-4	4FP21-3	5Z1-4
Needle jet	X-8	X-6	X-4
Needle position	4	3	4
Throttle valve	125	145	135
Pilot jet	20	17.5	42.5
Starter jet	25	40	30
Air screw (number of turns out) ...	1 to 1½	1 to 1½	1 to 1¼
Float height	26 ± 2.5 mm (1.047 ± 0.098 in)		32.0 ± 1 mm (1.26 ± 0.04 in)

Oil pump

Type	Trochoid
Inner rotor/outer rotor clearance	0.03–0.09 mm (0.0012–0.0035 in)
Side clearance	0.03–0.09 mm (0.0012–0.0035 in)

1 General description

The fuel system comprises a petrol tank, from which petrol is fed by gravity to the carburettors, via a diaphragm type fuel cock. There are three positions, ON, RESERVE and PRIMING. Before starting the engine, turn the fuel tap to the 'ON' position, this enables the fuel to flow to the carburettors when the engine has started.

If the fuel in the tank is too low to be fed to the carburettors in the ON position, turn the lever to RESERVE position, which provides a limited amount of additional fuel. Only when there is no fuel in the carburettors is it necessary to turn to the PRIMING position, which will allow fuel to flow to the carburettors even with the engine stopped. Once the engine has started be sure to return the lever to the ON position or RESERVE position.

The tap is operated by means of the induction vacuum in the right-hand inlet tract, to which the tap diaphragm is interconnected by a small bore hose.

Two Mikuni constant velocity carburettors are fitted to all models, interconnected by a throttle butterfly valve control rod and mounted as a unit on a common mounting bar. Each carburettor is provided with a starter system (choke) controlled by a plunger, the two being interconnected by a rod allowing simultaneous operation from a lever or push-pull knob at the left-hand carburettor. Two separate air filter units are fitted, one of which is hidden behind each frame side cover.

The lubrication system is fed by a trochoid type oil pump which is driven via an idler gear from the primary drive pinion. Oil picked up by the pump is strained by a gauze oil trap in the crankcase, where it is passed through a centrifugal oil filter to the working parts of the engine. A by-pass valve is included in the system, fitted within the oil filter centre bolt, which allows an uninterrupted flow of oil in the event of oil filter blockage. The gearbox and primary drive share the same oil as that used by the engine proper, the gearbox mainshaft being fed under pressure and the layshaft being lubricated by the gravity fed return system.

2 Petrol tank: removal and replacement

1   The fuel tank is retained at the forward end by two rubber buffers fitted either side of the underside of the tank which fit into cups on the frame top tube. The rear of the tank sits on a small rubber saddle placed on a frame cross-bar and is secured by a single bolt passing through a lug projecting from the tank.

2   To remove the tank, pull off the fuel line and vacuum tube at the petrol tap unions where they are held by spring clips. Raise the seat, remove the bolt, washer and damper from the rear of the tank, and lift the tank backwards and away from the machine.

3   When replacing the tank, reverse the above procedure. Make sure the tank seats correctly and does not trap any control cables or wires.

3 Petrol tap: removal and replacement

1   The petrol tap must be removed from the tank at regular intervals to gain access and allow cleaning of the filter gauze, which takes the form of a mesh pillar projecting into the petrol tank cavity. Before removing the tap, drain the fuel by attaching a suitable length of pipe to the tap union and turning the tap lever to the PRIME position, to allow an unrestricted flow of petrol. After draining the tank and detaching the vacuum pipe, remove the two screws which pass through the tap body flange and remove the tap.

2   The fuel filter should be cleaned thoroughly in petrol, agitating any adhering deposits with a soft brush. When replacing the tap note that there is an 'O' ring seal between the petrol tap body and the petrol tank, which must be renewed if it is damaged or if petrol leakage has occurred.

3   It is seldom necessary to remove the lever which operates the petrol tap, although occasions may occur when a leakage develops at the joint. Although the tank must be drained before the lever assembly can be removed, there is no need to disturb the body of the tap.

4   To dismantle the lever assembly, remove the two crosshead screws which pass through the plate on which the operating positions are inscribed. The plate can then be lifted away, followed by a spring, the lever itself and the seal behind the lever. The seal or lever 'O' ring will have to be renewed if leakage has occurred. Reassemble the tap in the reverse order. Gasket cement or any other sealing medium is NOT necessary to secure a petrol tight seal.

2.1 Petrol tank is retained at the rear by a single bolt

3.1 Withdraw the tap to gain access to the filter

3.4a Tap lever is secured by plate and wave washer

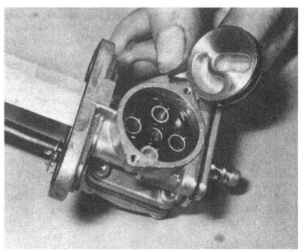

3.4b Lift out lever to gain access to seal and 'O' ring

4 Carburettors: removal

1   Remove the petrol tank as described in section 2 of this Chapter to gain access to the carburettors and then remove both frame side covers. Each side cover is a push fit at three locating points.

2   Loosen the screw clips which hold the carburettors to the inlet stubs, and the air hose unit to the twin air filter boxes and carburettor mouths. Each air filter box may be pulled from place after removing the retaining strap, which is secured at the lower end by a single bolt, and at the top by a hook arrangement. Disconnect the air hose breather tube from the breather cover union on the crankcase, after displacing the spring clip. The air hose unit can now be pulled from the carburettors and tilted backwards and downwards so that it clears the carburettors. Careful manipulation may be required as the clearance is limited.

3   Pull the carburettors back out of the inlet stubs and lift them out, towards the right-hand side of the machine as a unit. Support the carburettors and disconnect the throttle cable at the carburettor end. Lift the outer cable from the slotted abutment and displace the nipple and inner cable end from the throttle pulley.

5 Carburettors: dismantling and reassembly

1   In order to dismantle the carburettors they must first be removed from the mounting bracket to which they are each retained by two screws. The carburettors are also connected by a bracket which acts as the throttle cable holder. The bracket is retained by two of the four diaphragm cover screws on each carburettor top.

2   Commence by removing the right-hand carburettor from the mounting bracket. On carburettors which have a lever operated choke rod, remove the operating lever, which is held by a single screw. Slacken the grub screw holding each choke lifter fork and withdraw the rod to free the forks. Some carburettors have a two-position push-pull choke operating rod. The rod is located in any given position by a spring loaded steel ball in each carburettor, which locates with a series of depressions machined in the rod. To remove the rod, slacken the choke operating fork grub screws and withdraw the rod. As the rod end leaves the support lug in each carburettor, the steel ball and spring will tend to fly out. Arrangements should be made to prevent this happening, as the components are tiny and easily lost.

3   Slacken all four screws on each carburettor top, to prevent distortion, and then remove the inner four screws, to free the top bracket. Unscrew the two screws securing the right-hand carburettor to the mounting bracket and separate the two instruments at the interconnecting petrol transfer pipe. The mounting screws are often very tight and are prone to shearing. Great care should be exercised in their removal.

4   Invert each carburettor and remove the four screws that hold the float chamber to its base. Remove the hinge pin that locates the twin float assembly and lift the float from position. This will expose the float needle. The needle is very small and should be put away in a safe place so that it is not misplaced. Make sure that the float chamber gasket is in good condition. Do not disturb the gasket unless leakage has occurred or it appears damaged.

5   Check that the twin floats are in good condition and not punctured. Because they are made of brass it is possible to solder a damaged float. This form of repair should only be made in an emergency, when a set of new floats are not available. Soldering will affect the weight of the float assembly and result in a different petrol level.

XS250 and 360 models

6   The needle jet is a push fit in the base of the mixing chamber, being retained by a small 'O' ring. Check the needle jet

for wear together with the jet needle. After lengthy service, these two components should be renewed together, or high petrol consumption will result.

7   The float needle will also wear after lengthy service, and should be closely examined with a magnifying glass. Wear takes the form of a ridge or groove, which will cause the float needle to seat imperfectly. The needle and seating should always be renewed as a pair. The seating is a screw fit in the mixing chamber. Note the 'O' ring and also the tiny filter gauze, which is retained by the seat.

8   The main jet and pilot jet are both housed in the float chamber, The main jet is situated below a plug, which will unscrew from outside the float chamber. Always use a close fitting screwdriver when removing jets, or damage will result.

XS400 models

9   The jet configuration on XS400 model carburettors is slightly different from the other types in that all jets are located within the float chamber roof. Remarks on inspection and wear characteristics remain the same.

10   Unscrew the main jet from the centre turret and displace the needle jet from the body towards the venturi side of the carburettor. The pilot jet is closed by a brass plug, which may be removed to allow the jet to be cleaned. **Do not** remove the jet unless it is to be renewed as the method of removal which must be adopted will almost certainly enlarge the jet orifice. To remove the jet, insert a screwdriver and unscrew the jet until it can be felt to be free. The jet must now be unscrewed through the threads normally occupied by the blanking plug. In order to engage the jet with the second series of threads, it must be pulled and turned simultaneously. To accomplish this, select a short length of stiff wire whose diameter is slightly greater than the inside diameter of the jet bore. Taper the ends of the wire so that it may be inserted into the jet bore and pushed in firmly to engage the jet. The jet may now be pulled outwards and unscrewed.

11   The starter air jet is screwed into the left-hand side of the carburettor mouth and may be removed in the usual manner.

All models

12   Remove the two remaining screws which retain the carburettor top and lift the top from position, together with the piston spring. Carefully lift the diaphragm from position, bringing with it the piston and jet needle. Carefully check the condition of the diaphragm. If it has developed cracks or holes, it must be renewed as a unit, with the piston. The jet needle is retained by a nylon plate and is secured by a small circlip. The jet needle must be renewed if worn, as described in paragraph 6.

13   Before reassembly, clean the carburettors as described in the previous Section. The manually operated choke is unlikely to require attention throughout the normal service life of the machine.

14   Before the carburettors are reassembled, using the reversed dismantling procedure, each should be cleaned out thoroughly, using compressed air. Avoid using a piece of rag since there is always a risk of particles of lint obstructing the internal passageways or the jet orifices.

15   Never use a piece of wire or any pointed metal object to clear a blocked jet. It is only too easy to enlarge the jet under these circumstances and increase the rate of petrol consumption. If compressed air is not available, a blast of air from a tyre pump will usually suffice.

16   Do not use excessive force when reassembling a carburettor because it is easy to shear a jet or some of the smaller screws. Furthermore, the carburettors are cast in zinc-based alloy, which itself does not have a high tensile strength.

17   After replacing the carburettors on the machine they should be synchronised and adjusted, as described in Section 7 of this Chapter. Before refitting the air hose unit, check the synchronisation of the throttle valve butterflies.

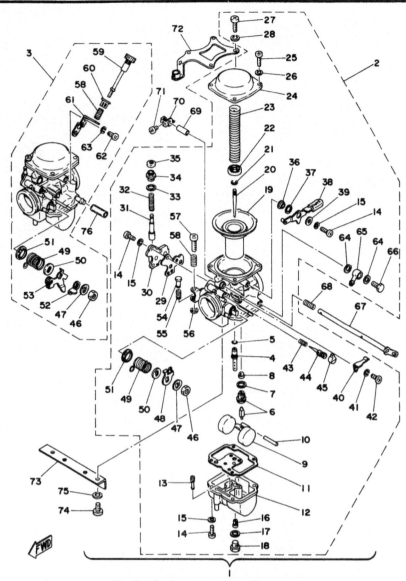

Fig. 2.1 Carburettor – lever choke type

1 Carburettors – complete	26 Spring washer – 4 off	51 Spacer – 2 off
2 L/H carburettor	27 Screw – 4 off	52 Throttle link arm
3 R/H carburettor	28 Spring washer – 4 off	53 Throttle control arm
4 Needle jet (main nozzle) – 2 off	29 Choke casing gasket – 2 off	54 Plunger
5 'O' ring – 2 off	30 Choke body – 2 off	55 Spring
6 Needle valve assembly – 2 off	31 Choke plunger – 2 off	56 'E' clip
7 Sealing washer – 2 off	32 Choke spring – 2 off	57 Throttle synchronisation screw
8 Filter screen – 2 off	33 Washer – 2 off	58 Spring – 2 off
9 Float assembly – 2 off	34 Housing – 2 off	59 Remote throttle stop screw
10 Float pivot pin – 2 off	35 Dust cap – 2 off	60 Bush
11 Gasket – 2 off	36 Collar	61 Bracket
12 Float bowl – 2 off	37 Washer	62 Screw – 2 off
13 Pilot jet – 2 off	38 Choke lever	63 Spring washer – 2 off
14 Screw – 15 off	39 Washer	64 Sealing washer – 4 off
15 Spring washer – 2 off	40 Spring plate	65 Banjo
16 Main jet – 2 off	41 Spring washer	66 Banjo bolt
17 Sealing washer – 2 off	42 Screw	67 Choke rod
18 Drain plug – 2 off	43 Spring – 2 off	68 Spring
19 Piston/diaphragm unit – 2 off	44 Pilot screw – 2 off	69 Collar
20 Needle – 2 off	45 Anti-tamper cap – 2 off	70 Choke fork – 2 off
21 Clip – 2 off	46 Nut – 2 off	71 Screw – 2 off
22 Needle seat – 2 off	47 Special washer – 2 off	72 Bracket
23 Piston spring – 2 off	48 Throttle arm	73 Mounting bar
24 Carburettor top – 2 off	49 Throttle return spring – 2 off	74 Screw – 4 off
25 Screw – 4 off	50 Collar – 2 off	75 Spring washer – 4 off
		76 Transfer pipe

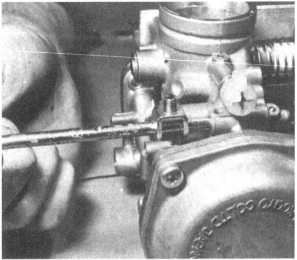

5.2a Loosen choke arm grub screws and withdraw the rod

5.2b Prevent detent ball and ...

5.2c ... spring from flying out

5.3 Each carburettor is held on mounting bar by two screws

5.4a Remove float bowl, held by four screws

5.4b Displace pivot pin to free floats and allow ...

5.4c ... removal of the float needle

5.4d Float valve seat unscrews for renewal or filter cleaning

5.10a Unscrew the main jet and ...

5.10b ... push out the needle jet

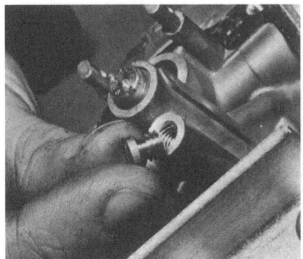

5.10c Pilot jet is enclosed by brass plug

5.10d Use tapered rod to withdraw the pilot jet

5.11 Starter jet screws into bell mouth

5.12a Remove the carburettor top and ...

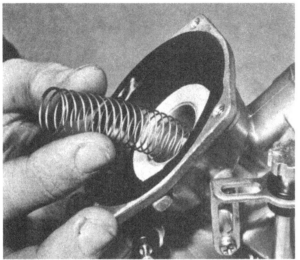

5.12b lift out the piston spring followed by ...

5.12c the piston/diaphragm unit

5.12d Remove the circlip to allow ...

5.12e withdrawal of the needle seat and ...

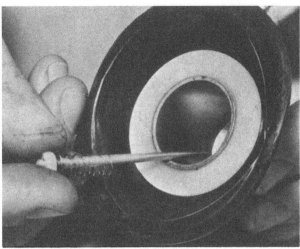

5.12f ... the piston needle, clip and spring

5.14a Diaphragm tab must locate with recess in body

5.14b Throttle link rod must be pushed between screw and plunger

6 Carburettors: adjusting float level

1   If problems are encountered with fuel overflowing from the float chambers, which cannot be traced to the float/needle assembly, or if consistent fuel starvation is encountered, the fault will probably lie in maladjustment of the float level. It will be necessary to remove the float chamber bowl from each carburettor to check the float level.

If the float level is correct the distance between the uppermost edge of the floats and the flange of the mixing chamber body will be as follows.

XS250 and 360	26 ± 2.5 mm (1.047 ± 0.098 in)
XS400	32 ± 1.0 mm (1.26 ± 0.040 in)

Adjustments are made by bending the float assembly tang (tongue), which engages with the float, in the direction required (see accompanying diagram).

7 Carburettors: synchronisation and adjustment

1   It is imperative that the two carburettors work in harmony with each other if maximum performance and fuel economy are

to be expected. The slow running mixture of each carburettor is set by the manufacturer to give the correct mixture and the cleanest exhaust emission. The manufacturers recommend that the mixture is not altered and to this end fit special plastic caps to the two pilot adjúster screws, to prevent movement of the screws except within a narrow margin. Synchronisation of the carburettors is therefore controlled by the adjuster screw on the throttle valve butterfly connecting arm.

2   Synchronisation of the butterfly valves can be made usually after detaching the airhose unit and then using a pair of vacuum gauges attached to the take-off unions provided, which project from the top of each inlet stub. When making the manual adjustment, rotate the adjuster screw as necessary, so that both butterfly valves open simultaneously and close simultaneously. After setting the carburettors initially, remove the vacuum hose from the right-hand take-off union and detach the sealing boot from the left-hand union. Connect up the two vacuum gauges as advised by the manufacturer of the gauge set. Start the engine and adjust the reading on right-hand gauge to within 5 Hg cm at 1,200 rpm. Adjustment is again made on the central throttle link screw. Synchronisation is now correct.

3   After completing synchronisation adjust the tick-over speed of the engine to 1,200 rpm by means of the remote throttle stop, which is located to the rear of the two carburettors, and which has a serrated nylon head.

4   Stop the engine and adjust the throttler cable so that there is 3-5 mm (0.12 - 0.20 in) slack, measured between the outer cable and the abutment on the carburettor bracket. Slack may be increased or reduced by means of the cable adjuster at the handlebar twist grip.

8 Carburettor settings

1   Some of the carburettor settings, such as the sizes of the needle jets, main jets and needle positions, etc, are predetermined by the manufacturer. Under normal circumstances it is unlikely that these settings will require modification, even though there is provision made. If a change appears necessary, it can often be attributed to a developing engine fault.

2   As stated previously, alteration of the idle mixture is not recommended. Some alterations however, can be made in the mid-range mixture by altering the height of the jet needle. This is accomplished by changing the position of the needle clip. Raising the needle will richen the mixture and lowering the needle will weaken the mixture.

3   Always err slightly on the side of a rich mixture, since a weak mixture will cause the engine to overheat. Reference to Chapter 3 will show how the condition of the spark plugs can be interpreted with some experience as a reliable guide to carburettor mixture strength. Flat spots in the carburation can usually be traced to a defective timing advancer. If the advancer action is suspect, it can be detected by checking the ignition timing with a stroboscope.

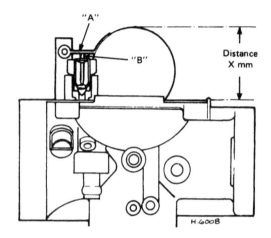

Fig. 2.2 Checking the float level height

9   Air filters: dismantling and cleaning

1   Two identical air filters are utilised, one of which is fitted behind each frame side cover. The two separate air filters are connected to the carburettors by an air hose unit which incorporates a balance pipe.
2   At regular intervals both air filters should be removed for cleaning in a similar manner. Remove both frame side covers, each of which has a three-point push fit method of retention.

7.3 Remote throttle stop screw positioned between carburettors

Slacken the air hose/air filter box screw clip and remove the single bolt holding the lower end of the retaining strap. Hinge the strap up and pull the complete air filter box from position. The air filter box is a two-piece moulding held by two screws. After removal of the screws, separate the two halves and withdraw the air filter element.
3   To clean the element, tap it lightly to loosen the accumulation of dust and then use a soft brush to sweep the dust away. Alternatively, compressed air can be blown into the element from the inside.
4   If the element is damp or oily it must be renewed. A damp or oily element will have a restrictive effect on the breathing of the carburettor and will almost certainly affect the engine performance.
5   On no account run without the air filters attached, or with the element missing. The jetting of the carburettors takes into account the presence of the air filters and engine performance will be seriously affected if this balance is upset.
6   To replace the element, reverse the dismantling procedure. Give a visual check to ensure that the inlet hoses are correctly located and not split or otherwise damaged. Check that the air filter cases are free from splits or cracks.

9.2a Air filter is fitted behind each frame side cover

9.2b Each filter box separates to allow element removal

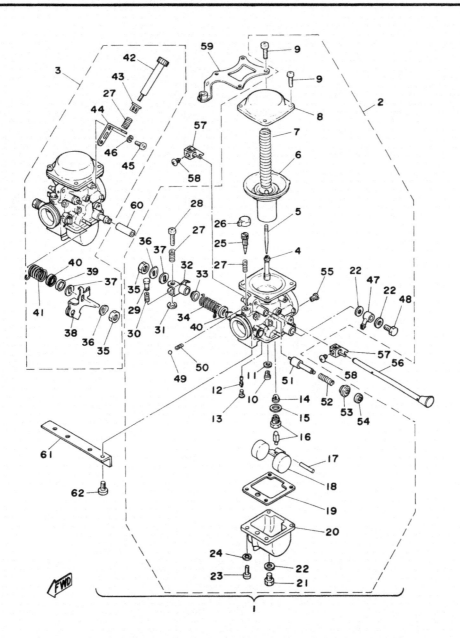

Fig. 2.3 Carburettor – two position choke type

1	Carburettors complete	22	Sealing washer – 4 off	43	Bush	
2	L/H carburettor	23	Screw – 8 off	44	Bracket	
3	R/H carburettor	24	Spring washer – 8 off	45	Screw – 2 off	
4	Needle jet (main nozzle) – 2 off	25	Pilot screw – 2 off	46	Spring washer – 2 off	
5	Needle – 2 off	26	Anti-tamper cap – 2 off	47	Banjo	
6	Piston/diaphragm unit – 2 off	27	Spring – 3 off	48	Banjo bolt	
7	Piston spring – 2 off	28	Throttle synchronisation screw	49	Steel ball – 2 off	
8	Carburettor top – 2 off	29	Plunger	50	Spring – 2 off	
9	Screw – 8 off	30	Spring	51	Choke plunger – 2 off	
10	Main jet – 2 off	31	E clip	52	Plunger spring – 2 off	
11	Washer – 2 off	32	Throttle link arm	53	Cap – 2 off	
12	Pilot jet – 2 off	33	Bush	54	Cover – 2 off	
13	Plug – 2 off	34	LH throttle return spring	55	Pilot air jet – 2 off	
14	Filter screen – 2 off	35	Nut – 2 off	56	Choke operating rod	
15	Sealing washer – 2 off	36	Special washer – 2 off	57	Choke fork – 2 off	
16	Needle valve assembly – 2 off	37	Special washer – 2 off	58	Screw – 2 off	
17	Float pivot pin – 2 off	38	Throttle control lever	59	Bracket	
18	Float assembly – 2 off	39	Spacer	60	Transfer pipe	
19	Gasket – 2 off	40	Seal – 2 off	61	Mounting bar	
20	Float bowl – 2 off	41	R/H throttle return spring	62	Screw – 4 off	
21	Drain plug – 2 off	42	Remote throttle stop screw			

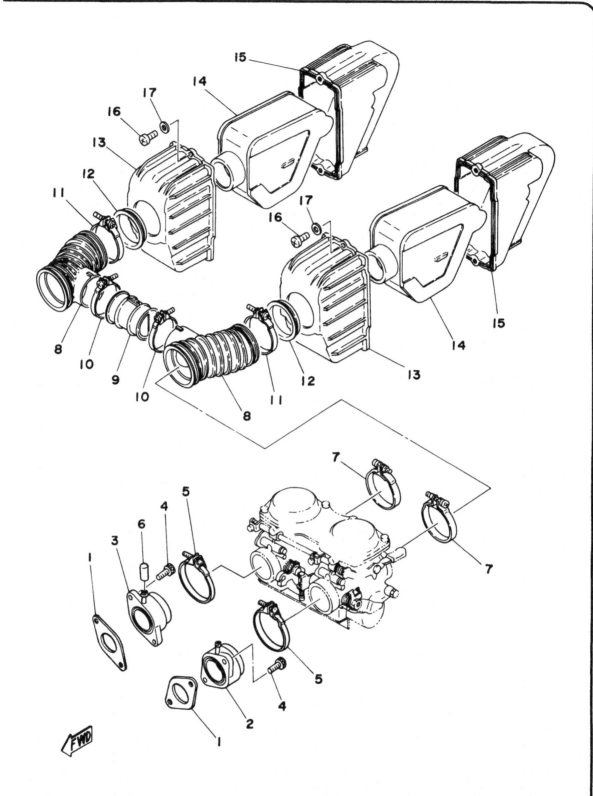

Fig. 2.4 Air filter assembly

1	Gasket – 2 off	7	Screw clip – 2 off	13	Air filter casing half – 2 off
2	Inlet stub	8	Inlet hose – 2 off	14	Air filter element – 2 off
3	Inlet stub	9	Balance pipe	15	Air filter casing half – 2 off
4	Socket screw – 4 off	10	Screw clip – 2 off	16	Screw – 4 off
5	Screw clip – 2 off	11	Screw clip – 2 off	17	Plain washer – 4 off
6	Dust excluder	12	Grommet – 2 off		

10 Exhaust system

1   Unlike a two-stroke, the exhaust system does not require such frequent attention because the exhaust gases are usually of a less oily nature.
2   Do not run the machine with the exhaust baffles removed, or with a quite different type of silencer fitted. The standard production silencers have been designed to give the best possible performance, whilst subduing the exhaust note to an acceptable level. Although a modified exhaust system or one without baffles, may give the illusion of greater speed as a result of the changed exhaust note, the chances are that performance will have suffered accordingly.

11 Oil pump: removal, examination and replacement.

1   The oil pump is contained within the primary drive casing where it is driven via an idler gear from the primary drive pinion. The oil pump may be removed whilst the engine is still in the frame, after the oil has been drained and the primary drive cover removed.
2   The oil pump is retained by three socket bolts and is located by a single dowel pin, to allow adjustment of the primary drive pinion/idler pinion backlash. Remove the three bolts and lift the pump away, complete with the idler pinion. Slide the pinion off the stub spindle which projects from the rear of the pump unit.
3   Commence dismantling by removing the three countersunk screws which secure the pump cover plate. The screws will be very tight and require careful loosening to avoid damage to the heads. After removal of the cover, lift out the outer rotor. The inner rotor shaft and oil pump pinion are integral units. No provision is made for their removal from the pump body after initial assembly.
4   Incorporated in the pump casting is the pressure release valve which allows control of the feed oil pressure. If excess pressure is created, the valve plunger lifts, allowing oil to flow directly to the sump. The valve is comprised of a plunger, spring and spring seat retained by a split pin. Straighten the split pin and depress the spring seat slightly with a screwdriver. Withdraw the split pin whilst maintaining the pressure on the spring

seat. This action will prevent the spring and seat from flying out. Remove the seat, spring and plunger.
5   Wash all the pump components with petrol and allow them to dry before carrying out an examination. Before partially reassembling the pump for various measurements to be carried out, check the casting for breakage or fracture, or scoring on the inside perimeter.
6   Refit the outer rotor and measure the clearance between the tip of the inner rotor and a peak on the outer rotor. The clearance should be 0.03 - 0.09 mm (0.001 - 0.002 in). Check also the clearance between the side of the adjacent peak and tip. The clearance should be the same. If wear is evident, the pump should be renewed as a unit. Place a straight edge across the pump body face and check the clearance between the rotor faces and the straight edge. The side float should not exceed 0.10 - 0.18 in (0.004 - 0.007 in). Excessive wear indicates need for renewal.
7   Examine the rotors and the pump body for signs of scoring, chipping or other surface damage which will occur if metallic particles find their way into the oil pump assembly. Renewal of the pump is the only remedy under these circumstances.
8   Reassemble the pump and the pump casing by reversing the dismantling procedure. The pump components must be ABSOLUTELY clean before assembly is commenced. The outer rotor should be fitted with the small punch mark on one face pointing inwards. Make sure all parts of the pump are well lubricated before the outer cover is replaced and that there is plenty of oil between the inner and outer rotors. Tighten the end cover down evenly and continually check the drive pinion revolves freely. A stiff pump is usually due to dirt on the rotor faces.
9   Fit the single location dowel into the casing and replace the pump complete with the idler pinion. Insert the three socket bolts and tighten them until it is just possible to move the pump up and down in line with the vertical plane of the engine. The backlash between the primary drive pinion and the oil pump idler pinion must now be adjusted. Select a 0.4 mm (0.015 in) feeler gauge and slide it between the driven sides of the central meshed teeth of the two pinions. The position of the feeler gauge will be just to the right of an imaginary line drawn between the centres of the two pinions. Push the oil pump forward bracket arm upwards so that the feeler gauge is gripped lightly and then tighten the socket bolt. Tighten the remaining two bolts and recheck the backlash.

11.3 The oil pump side plate is retained by three countersunk screws

11.4a Displace the split pin to allow ...

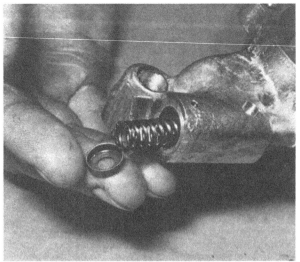

11.4b ... removal of the oil pressure release spring and ...

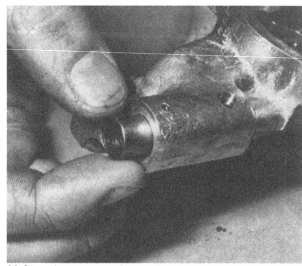

11.4c ... the pressure release valve plunger

11.6a Check the outer rotor/housing clearance and ...

11.6b ... check the inner rotor/outer rotor tip clearance

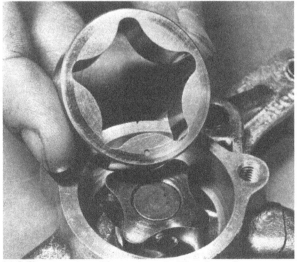

11.8 Fit outer rotor with punch marked face inwards

11.9 Pump idler gear spindle projects from the oil pump body

12 Oil filter: renewing the element

1   The oil filter is contained within a semi-isolated chamber at the front of the crankcase. Access to the element is made by unscrewing the filter cover centre bolt, which will bring with it the cover and also the element. Before removing the cover, place a receptacle beneath the engine to catch the engine oil contained in the filter chamber.

2   When renewing the filter element it is wise to renew the filter cover 'O' ring at the same time. This will obviate the possibility of any oil leaks. Do not overtighten the centre bolt on replacement; the correct torque setting is 1.3 - 1.7 kg m (9.5 - 12.0 lb ft).

3   The filter by-pass valve, comprising a plunger and spring, is situated in the bore of the filter cover centre bolt. It is recommended that the by-pass valve be checked for free movement during every filter change. The spring and plunger are retained by a pin across the centre bolt. Knocking the pin out will allow the spring and plunger to be removed for cleaning.

4   Never run the engine without the filter element or increase the period between the recommended oil changes or oil filter changes. Engine oil should be changed every 2,000 miles and the element changed every 4,000 miles. Use only the recommended viscosity.

13 Oil pressure warning lamp

1   An oil pressure warning lamp is incorporated in the lubrication system to give immediate warning of excessively low oil pressure.

2   The oil pressure switch is screwed into the crankcase, directly behind the final drive sprocket cover and is connected with a warning light in the lighting panel on the handlebars. The light should be on whenever the ignition is on but will usually go out at about 1,500 rpm.

3   If the oil warning lamp comes on whilst the machine is being ridden, the engine should be switched off immediately, otherwise there is a risk of severe engine damage due to lubrication failure. The fault must be located and rectified before the engine is re-started and run, even for a brief moment. Machines fitted with plain shell bearings rely on high oil pressure to maintain a thin oil film between the bearing surfaces. Failure of the oil pressure will cause the working surfaces to come into direct contact, causing overheating and eventual seizure.

12.1a Oil filter chamber is secured to front of crankcase by ...

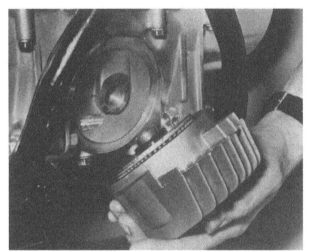

12.1b ... a hollow bolt containing the by-pass valve

12.1c Do not omit oil filter spring and washer on reassembly

14 Fault diagnosis: fuel system

Symptom	Cause	Remedy
Engine gradually fades and stops	Fuel starvation	Check vent hole in filler cap. Sediment in filter bowl or float chamber. Dismantle and clean.
Engine runs badly. Black smoke from exhausts	Carburettor flooding	Dismantle and clean carburettor. Check for punctured float or sticking float needle
Engine lacks response and overheats	Weak mixture Air cleaner disconnected or hose split Modified silencer has upset carburation	Check for partial block in carburettors. Reconnect or renew hose. Replace with original design.
Oil pressure warning light comes on	Lubrication system failure	Stop engine immediately. Trace and rectify fault before re-starting.
Engine gets noisy	Failure to change engine oil when recommended	Drain off old oil and refill with new oil of correct grade. Renew oil filter element.

Chapter 3 Ignition System

For information relating to later models, see Chapter 7

Contents

Specifications

Ignition timing (BTDC)

Retarded	$10°$
Advanced	$36°$
Contact breaker gap	0.30–0.40 mm (0.012–0.016 in)
Dwell angle	$105°$

Ignition coil

Primary winding resistance ...	4.0 ohms ± 10% at 20°C (68°F)
Secondary winding resistance ...	9.5 K ohms ± 20% at 20°C (68°F)

Sparking plugs

	XS250	XS360	XS400
NGK	BP-7ES	BP-6ES	BP-7ES
N-D	W22EP	W20EP	W22EP
Motorcraft	AG12	AG22	AG12
Plug gap	0.7–0.8 mm (0.028–0.032 in)		

1 General description

1 The Yamaha XS250, 360 and 400 twins are fitted with a 12 volt negative earth electrical system. Current derived from the crankshaft mounted alternator supplies twin ignition coils separately located which, in conjunction with a twin contact breaker assembly, provide the necessary spark at the correct time, to ignite the mixture in the cylinders.

2 The contact breakers are fitted within a chamber in the left-hand side of the cylinder head, where the cam and automatic timing unit are driven from the end of the camshaft.

3 Ignition source power is fed from the battery to the ignition coil primary windings. When the contact breaker opens, the low tension circuit is interrupted, and a high voltage is produced in the ignition coil secondary windings by magnetic induction. The high voltage passes through an HT lead to the spark plug.

4 A condenser in the contact breaker circuit prevents the contact breaker points from arcing as they open and close, to ensure that the spark at the spark plug gap has full intensity. The condenser is a twin unit contained within a single canister mounted on the frame top-tube or on the left-hand side of the battery box.

2 Charging system: checking the output

The charging system, which includes the alternator, rectifier and voltage regulator, can be checked satisfactorily only by using test equipment of the multi-meter type. If the charging performance is suspect, some initial tests may be made to determine the condition of the various components, as described in Chapter 6.

3 Contact breaker: adjustment

1 To gain access to the contact breaker assembly, it is necessary to remove the cover plate which is held by two cross head screws to the cylinder head. Note that the cover has a paper gasket to prevent the ingress of water.

2 Remove the alternator inspection cover and rotate the engine until the left-hand points are fully open. Removal of the spark plugs will aid rotation. Examine the faces of the points. If they are blackened and burnt, or badly pitted, it will be necessary to remove them for further attention. See Section 4 of this Chapter.

3   Adjustment is effected by slackening the two screws through the plate of the fixed contact breaker point and moving the point either closer, or further away, from the moving contact until the gap is correct as measured by a feeler gauge. The correct gap with the points FULLY OPEN is 0.3 - 0.4 mm (0.012 - 0.016 in).

4   Two small projections on the contact breaker base plate permit the insertion of a screwdriver to lever the adjustable point into its correct location. Repeat this operation if there is any doubt about the accuracy of the measurement. Although the adjustment is relatively easy, it is of prime importance.

5   Repeat the same procedure for the right-hand set of contact breakers, after turning the engine again so that the points are fully open.

6   Before replacing the cover and gasket, place a few drops of thin oil on the cam lubrication wick. Do not over lubricate, as excess oil will eventually find its way onto the two contact points, causing the ignition circuit to malfunction.

4   Contact breaker points: removal, renovation and replacement

1   If the contact breaker points are burned, pitted or badly worn, they should be removed for dressing. If it is necessary to remove a substantial amount of material before the faces can be restored, the points should be renewed.

2   To remove the contact breaker points, unclip the contact breaker leads at their snap connectors. Remove the four slot headed screws thereby releasing both the contact breakers. **Do not unscrew the two crosshead screws** that retain the contact breaker back plate, otherwise the ignition will have to be retimed. The contact breakers can themselves be separated by removing their circlip and washers and pulling the movable point off the pin, to facilitate inspection and cleaning.

3   The points should be dressed with an oil stone or fine emery cloth. Keep them absolutely square throughout the dressing operation, otherwise they will make angular contact on reassembly, and rapidly burn away. If emery cloth is used, it should be backed by a flat strip of steel. This will reduce the risk of rounding the edges of the points.

4   Replace the contacts by reversing the dismantling procedure, making quite certain that the insulating washers are fitted in the correct way. In order for the ignition system to function at all, the moving contact and the low-tension lead must be perfectly insulated from the base plate and fixed contact. It is advantageous to apply a very light smear of grease to the pivot pin, prior to replacement of the moving contact.

5   Check, and if necessary, re-adjust the contact breaker points when they are in the fully open position.

5   Ignition timing: checking and re-setting

1   It is recommended that the ignition timing check be carried out statically (with the engine stopped), rather than dynamically using a stroboscope. Although use of a stroboscope will enable checking of the full advance operation, it should be noted that in running the engine, oil will issue from the open crankcase left-hand half with the resultant loss of engine lubrication. To carry out the static timing check proceed as follows.

2   Start by inspection and setting the contact breakers as described previously. Either drain the engine oil, or prepare to catch oil spillage when the circular inspection cover on the left-hand engine side (4 screws) is removed.

3   With the cover removed, turn the engine slowly and look through the inspection slot in the casing upper section to identify the timing marks on the alternator rotor periphery. For each cylinder there will be a full advance mark (two single lines close together), followed by the timing mark (LF for left cylinder, RF for right), and then the TDC mark (LT for left cylinder, RT for right). For the static timing check, the LF (left cylinder) and RF (right cylinder) marks are used. Note that the actual mark for alignment purposes is the full scribed line next to the LF or RF (see photo 5.1a).

4   Check the left-hand cylinder first. In order to determine the exact moment of contact breaker point separation connect up an audible points checker, multimeter set to the resistance range or a battery and bulb test circuit across the left-hand contact breaker set. This is best done by connecting the equipment between the orange wire from the points and a good earth on the engine. When the points separate the buzzer will sound, meter needle swing or bulb light (as applicable).

5   Rotate the engine slowly anticlockwise until the left cylinder is just approaching the LF mark on its compression stroke. If the timing is set correctly, the points should separate when the LF mark aligns with the static casing mark at the top of the inspection aperture.

6   If the timing is incorrect, align the rotor LF mark with the static casing mark, then slacken the two screws which clamp the contact breaker baseplate to the cylinder head. Rotate the complete contact breaker assembly until the equipment indicates that the points have just separated and tighten the baseplate screws. Recheck the timing by rotating the engine backwards about 45° and then forwards again, until the LF mark aligns precisely with the index mark. The engine must be rotated one way and then the other to take up any backlash in the timing chain (camshaft chain).

7   Repeat the timing check procedure on the right-hand contact-breaker, using the RF mark on the alternator rotor and the grey wire from the right-hand contact breaker set. If the timing

3.3 A = Points adjustment screws, LH cylinder, B = adjustment screws, RH cylinder

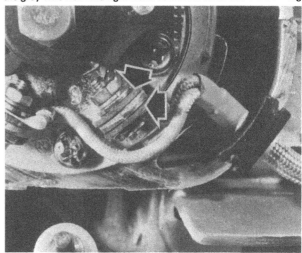

4.4 Ensure insulation washers are fitted in correct sequence

Electrode gap check - use a wire type gauge for best results

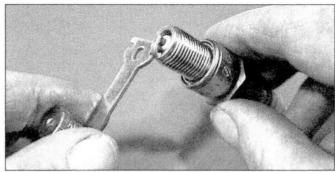

Electrode gap adjustment - bend the side electrode using the correct tool

Normal condition - A brown, tan or grey firing end indicates that the engine is in good condition and that the plug type is correct

Ash deposits - Light brown deposits encrusted on the electrodes and insulator, leading to misfire and hesitation. Caused by excessive amounts of oil in the combustion chamber or poor quality fuel/oil

Carbon fouling - Dry, black sooty deposits leading to misfire and weak spark. Caused by an over-rich fuel/air mixture, faulty choke operation or blocked air filter

Oil fouling - Wet oily deposits leading to misfire and weak spark. Caused by oil leakage past piston rings or valve guides (4-stroke engine), or excess lubricant (2-stroke engine)

Overheating - A blistered white insulator and glazed electrodes. Caused by ignition system fault, incorrect fuel, or cooling system fault

Worn plug - Worn electrodes will cause poor starting in damp or cold weather and will also waste fuel

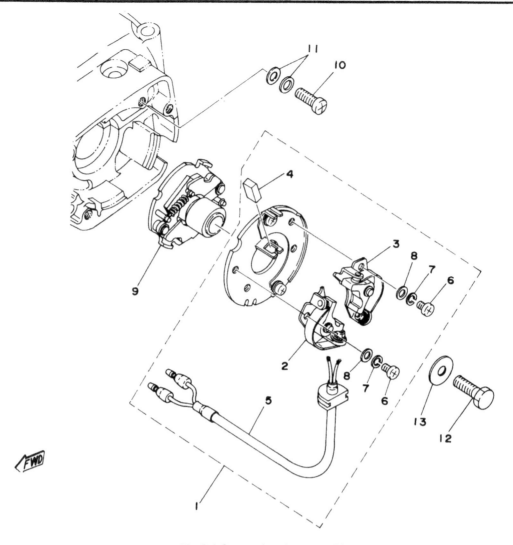

Fig. 3.1 Contact breaker assembly

1	Contact breaker plate assembly	6	Screw – 4 off	10	Screw – 2 off
2	Contact breaker assembly – left	7	Spring washer – 4 off	11	Washer – 4 off
3	Contact breaker assembly – right	8	Plain washer – 4 off	12	Bolt
4	Lubricator wick	9	Automatic timing unit	13	Washer
5	Lead wire assembly				

5.1a Timing mark for left-hand cylinder together with ...

5.1b ... index pointer on casing allow ease of ignition timing

5.3 LH cylinder timing screw (A), RH cylinder timing screw (B)

is incorrect **DO NOT** slacken the two screws previously used for timing. The right-hand contact breaker is mounted on a separate smaller base plate, retained by two screws passing through elongated holes in the plate. The elongated holes allow a limited amount of plate movement for ignition timing.

8   Disconnect the test circuit from the points and refit all disturbed components. Replenish any lost engine oil.

6   Condenser: function, removal and replacement

1   The condenser unit, which is located on the top tube of the frame, or to the left of the battery box, is comprised of two condensers within a common body, the body forming the common earth lead. Therefore good earthing of the mounting bracket is essential.

2   The condenser has two functions. Firstly it reduces the sparking at the contacts (and hence prevents the rapid wear of the points). Its second and most important function is to greatly increase the induced voltage in the secondary windings and hence strength of the high tension spark at break. Without the condenser the spark would be very weak.

3   If the engine is difficult to start, or if misfiring occurs, it is possible that a condenser is at fault. To check whether a condenser has failed, observe the points whilst the engine is running, after removing the contact breaker cover. If excessive sparking occurs across one set of points, and they have a blackened or burnt appearance, it may be assumed the condenser in that circuit is no longer serviceable. It follows that if one condenser fails, the complete unit must be renewed.

4   Access to the condenser can be made after removal of the petrol tank (see Chapter 2, Section 2) or by raising the dualseat, depending on the model and the location of the condenser. Disconnect the condenser leads at the block connector and remove the single retaining screw.

7   Condensers: testing

1   Without the appropriate test equipment, there is no alternative means of verifying whether a condenser is still serviceable.

2   Bearing in mind the low cost of a condenser, it is far more satisfactory to check whether it is malfunctioning by direct replacement.

8   Ignition coils: checking

1   Each cylinder has its own ignition circuit and if one cylinder misfires, one half of the complete ignition system can be eliminated immediately. The components most likely to fail in the circuit that is defective are the condenser and the ignition

coil since contact breaker faults should be obvious on close examination. Replacement of the existing condenser will show whether the condenser is at fault, leaving by the process of elimination the ignition coil.

2   The ignition coil can best be checked using a multimeter set to the resistance position. Detach the orange lead and red/white lead at their snap connectors and detach the spark plug cap from the spark plug. Measure the primary winding resistance and the secondary winding resistance by connecting the multimeter as shown in the accompanying diagram.

The resistance values for each circuit should be as follows:

Primary coil resistance   4.0 ohms $\pm$ 10% at 20°C
Secondary coil resistance 9.5K ohms $\pm$ 20% at 20°C

Slight variation may be encountered if the ambient temperature departs greatly from that given. If the values differ from those given, the coil is faulty.

3   If the multimeter is not available, and by means of testing, the other components have been found to be satisfactory, the following method may be used to give an estimation of the coil's condition. Remove the suppressor cap and bare the inner wire. Remove the contact breaker cover and turn the engine over until the contact breaker points relevant to the coil to be tested are closed. Turn the ignition on and using an insulated screwdriver flick the points open and shut. If the bared end of the HT lead is held approximately 6 mm from an earthing point (the cylinder head) whilst this is done, a blue spark should jump the gap. If the spark is unable to jump a gap, or is yellowish in colour, the coil is probably at fault.

4   The ignition coils are sealed units and it is not possible to effect a satisfactory repair in the event of failure. A new coil must be fitted.

5   The ignition coils are mounted as a pair underneath the petrol tank. They bolt direct to metal plates on the duplex top frame tube and face in a rearward direction, parallel to the axis of the machine.

9   Automatic timing unit: examination

1   The automatic timing mechanism rarely requires attention, although it is advisable to examine it periodically, when the contact breaker is receiving attention. It is retained by a small bolt and washer through the centre of the integral contact breaker cam and can be pulled off the end of the camshaft when the contact breaker plate is removed.

2   The unit comprises spring loaded balance weights, which move outward against the spring tension as centrifugal force increases. The balance weights must move freely on their pivots and be rust-free. The tension springs must also be in good condition. Keep the pivots lubricated and make sure the balance weights move easily, without binding. Most problems arise as a result of condensation, within the engine, which causes the unit to rust and balance weight movement to be restricted.

3   The automatic timing unit mechanism is fixed in relation to the crankshaft by means of a dowel. In consequence the mechanism cannot be replaced in anything other than the correct position. This ensures accuracy of ignition timing to within close limits, although a check should always be made when reassembly of the contact breakers is complete.

4   The correct functioning of the auto-advance unit can be checked when the engine is running by the use of a stroboscopic light. If a strobe light is available, connect it to the ignition circuit as directed by the manufacturer of the light. With the engine running, direct the beam of the light at the index mark on the alternator cover, and so that the alternator rim is illuminated. Check that the fully retarded timing mark (either LF or RF depending on which contact breaker is connected to the light) is in alignment with the pointer with the engine at 1,200 rpm tickover. Increase the engine speed to 3000 rpm. As the speed increases through this range the firing mark should move smoothly away from the index mark to be replaced by the unmarked scribed advance marks. Sluggishness or erratic movement indicates a poorly performing ATU.

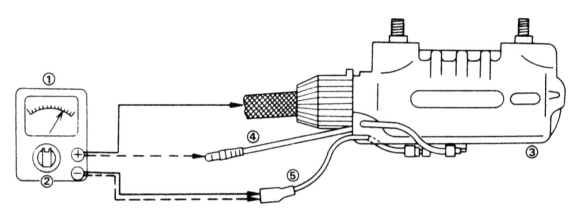

Fig. 3.2 Testing ignition coil continuity

1 *Multimeter*
2 *Set tester on the 'Resistance' position*
3 *Ignition coil*
4 *Red/white*
5 *Orange*

8.4 Ignition coils are attached to frame tubes, below petrol tank

9.1 The ATU is fitted to the camshaft left-hand end

9.2 Check the condition of the ATU bob-weights and springs

10 Spark plugs: checking and resetting the gaps

1   The XS250 and 400 models are fitted as standard with NGK BP-7ES or Nippon Denso W22EP spark plugs. The XS2360 model has either NGK BP-6ES or Nippon Denso ND W20EP spark plugs. All plugs are gapped within the range 0.7-0.8 mm (0.028 - 0.032 in). Certain operating conditions may indicate a change in spark plug grade, the type recommended by the manufacturer gives the best, allround service.

2   Check the gap of the plug points during every three monthly or 2000 mile service. To reset the gap, bend the outer electrode to bring it closer to the centre electrode and check that a 0.7 mm (0.028 in) feeler gauge can be inserted. Never bend the central electrode or the insulator will crack, causing engine damage if the particles fall in whilst the engine is running.

3   With some experience, the condition of the spark plug electrodes and insulator can be used as a reliable guide to engine operating conditions. See accompanying photographs.

4   Beware of overtightening the spark plugs, otherwise there is risk of stripping the threads from the aluminium alloy cylinder heads. The plugs should be sufficiently tight to sit firmly on their copper sealing washers, and no more. Use a spanner which is a good fit to prevent the spanner from slipping and breaking the insulator.

5   If the threads in the cylinder head strip as a result of over tightening the spark plugs, it is possible to reclaim the head by the use of a Helicoil thread insert. This is a cheap and convenient method of replacing the threads; most motorcycle dealers operate a service of this kind.

6   Make sure the plug insulating caps are a good fit and have their rubber seats. They should also be kept clean to prevent tracking. These caps contain the suppressors that eliminate both radio and TV interference.

11 Fault diagnosis: ignition system

Symptom	Cause	Remedy
Engine will not start	Faulty ignition switch	Operate switch several times in case contacts are dirty. If lights and other electrics function, switch may need renewal.
	Starter motor not working	Discharged battery. Use kickstart until battery is recharged.
	Short circuit in wiring	Check whether fuse is intact. Eliminate fault before switching on again.
	Completely discharged battery	If lights do not work, remove battery and recharge.
Engine misfires	Faulty condenser in ignition circuit	Renew condenser and re-test.
	Fouled spark plug	Renew plug and have original cleaned.
	Poor spark due to generator failure and discharged battery	Check output from generator. Remove and recharge battery.
Engine lacks power and overheats	Retarded ignition timing	Check timing and also contact breaker gap. Check whether auto-timing unit has jammed.
Engine 'fades' when under load	Pre-ignition	Check grade of plugs fitted; use recommended grade only.

Chapter 4 Frame and Forks

For information relating to later models, see Chapter 7

Contents

Specifications

Front forks

Type	Telescopic, hydraulically damped
Damping fluid capacity	130 cc (4.4/3.7 US/Imp fl oz)
Fluid specification	SAE 10W/30 engine oil or a fork oil
Fork spring free length	484 mm (19.05 in)

Rear suspension

Type	Swinging arm on two hydraulically damped suspension units
Spring length	205 mm (8.1 in)
Swinging arm free play (maximum)	1 mm (0.04 in)

1 General description

The Yamaha XS250, 360 and 400 models share a similar frame of traditional design, constructed from welded tubular members and having all main tubes, except the downtube, duplicated for strength.

The front forks are of the conventional telescopic type where the fork springs are contained within the fork stanchions and an oil damping medium is used. Rear suspension is of the swinging arm type, using oil filled suspension units to provide the necessary damping action. The units are adjustable so that the spring ratings can be effectively changed within certain limits, to match the load carried.

2 Front forks: removal from the frame

1   It is unlikely that the front forks will have to be removed from the frame as a complete unit, unless the steering head assembly requires attention or if the machine suffers severe frontal damage. Removal of the individual fork legs for inspection and renovation can be accomplished easily, without the need for disturbing the fork yokes and steering head bearings. If required, removal of the fork legs as described in this section may be followed by steering yoke and bearing dismantling, refering to the procedure in the following section.

2   Place the machine on the centre stand so that it rests securely on level ground. Raise the front wheel well clear of the ground by placing wood blocks below the crankcase.

3   Remove the front wheel as described in Chapter 5 Section 4.

4   On disc brake models detach the front brake caliper unit from the fork leg by removing the two bolts which pass through the fork leg lugs into the caliper support bracket. Release the hydraulic hose by removing the hose clamp bolted to the mudguard. **Do not** disconnect the hose from the caliper as fluid loss will result, and the front brake system will have to be bled of air. Swing the caliper unit back out of the way and tie it to a suitable part of the frame.

5   Remove the front mudguard. It is held in place by two bolts passing into each fork leg. Slacken the pinch bolts on the fork upper yoke and lower yoke which clamps the fork legs in place. Each fork leg may be pulled down and out of the fork yokes as a complete unit. It may be necessary to spring the clamps apart with a screwdriver blade to release the grip on the fork stanchions. Care should be taken if this method is adopted because excess force will fracture the yoke casting.

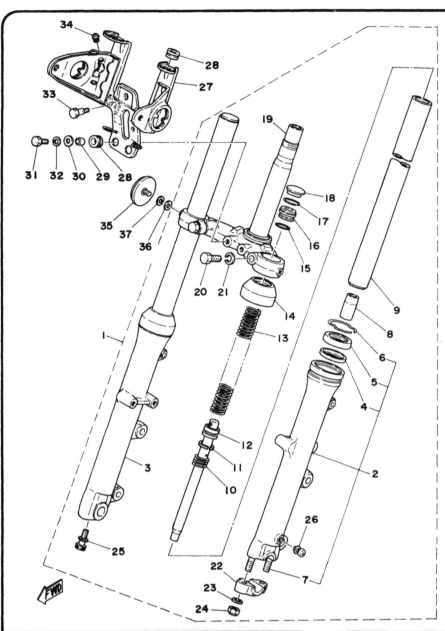

Fig. 4.1 Front fork assembly

1 Front fork assembly
2 Left-hand lower fork leg
3 Right-hand lower fork leg
4 Oil seal spacer – 2 off
5 Oil seal – 2 off
6 'E' clip – 2 off
7 Stud – 2 off
8 Damper rod seat – 2 off
9 Fork stanchion (inner tube) – 2 off
10 Rebound spring – 2 off
11 Piston ring – 2 off
12 Fork stanchion (upper tube) – 2 off
13 Fork spring – 2 off
14 Dust seal – 2 off
15 'O' ring – 2 off
16 Spring retaining plug – 2 off
17 Circlip – 2 off
18 Front fork cap – 2 off
19 Steering stem and bottom yoke
20 Pinch bolt – 2 off
21 Spring washer – 2 off
22 Spindle clamp
23 Plain washer – 2 off
24 Nut – 2 off
25 Clamp bolt – 2 off
26 Drain plug – 2 off
27 Headlamp bracket
28 Grommet – 4 off
29 Spacer – 2 off
30 Plain washer – 2 off
31 Bolt – 2 off
32 Spring washer – 2 off
33 Bolt
34 Rubber plug – 4 off
35 Reflector – 2 off
36 Plain washer – 2 off
37 Spring washer

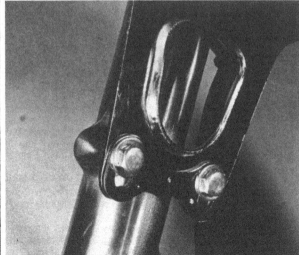

2.5a The mudguard is held by two bolts on each side

2.5b Slacken the upper pinch bolt and ...

2.5c ... the lower pinch bolt and ...

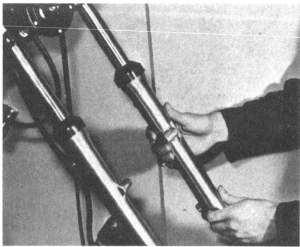

2.5d ... draw the fork leg downwards as a unit

3 Steering head yokes and bearings: removal from the frame

1   After removal of the fork legs as described in the preceding section, the steering head yokes and bearings may be detached to complete fork removal.
2   Raise the dualseat and disconnect the battery positive (+) lead. This will isolate the electrical system and prevent accidental shorting when the controls and headlamp leads are disconnected.
3   On disc brake models, disconnect the front brake stop lamp switch leads and remove the two bolts holding the master cylinder/reservoir unit to the handlebars. The master cylinder and caliper unit to which it is still interconnected by the brake hose may be removed from the machine. Do not allow the master cylinder to be inverted. This removal procedure precludes the need for bleeding the brake system on reassembly. On drum brake models, disconnect the brake cable from the handlebar lever.
4   Detach the controls from the handlebars, disconnecting cables and electrical leads where required. The electrical connections are made by block or snap connectors and the wires are colour coded to aid reassembly. Remove the handlebars after detaching the two clamps, held by two bolts each.
5   The speedometer and tachometer unit is secured on a mounting plate which is held on the fork upper yoke by two studs and nuts. Detach the instrument drive cables and remove the securing nuts, to allow the complete unit to be lifted away from the machine.
6   The headlamp unit is mounted in a bracket secured by two bolts to the lower yoke. The upper part of the bracket locates on studs. Remove the two bolts, lower the bracket slightly to disengage the upper mountings, and lift the assembly away from the yokes.
7   Remove the crown bolt and washer from the centre of the upper yoke. After slackening the pinch bolt at the rear of the yoke, the yoke may be lifted upwards, off the steering stem.
8   To release the lower yoke and steering head stem, unscrew the slotted sleeve nut at the top of the steering stem. A suitable 'C' spanner is provided in the tool kit with which to loosen the sleeve nut, but a soft punch can be used in the absence of the correct tool. As the steering head is released, the uncaged ball bearings from the lower race will be released and care should be taken to catch them by wrapping a rag around the bearing area. The bearings in the upper race will almost certainly remain in position.

4 Front forks: dismantling

1   It is advisable to dismantle each fork leg separately, using an identical procedure. There is less chance of unwittingly exchanging parts if this approach is adopted. Commence by draining each fork leg of damping oil; there is a drain plug in each lower leg above and to the rear of each wheel spindle housing.
2   Place the fork stanchion in a vice fitted with soft jaws, with a length of rubber inner tube around the stanchion, to prevent damage. Prise the rubber plug from the top of the stanchion. The fork spring is retained by a close fitting plug secured by an internal circlip within the stanchion end. To remove the circlip, the plug must be depressed slightly against the spring pressure and the circlip prised out with a small screwdriver. The help of a second person should be enlisted to depress the plug. After removal of the circlip, release the pressure and remove the plug, followed by the fork spring.
3   Place the fork lower leg in the vice and unscrew the socket screw recessed into the housing which carries the front wheel spindle. Prise the dust excluder from position and slide it up the fork stanchion. The stanchion can be pulled out of the lower fork leg. Remove the damper rod seat. Invert the upper tube and push the damper rod out of position towards the top end of the tube.
4   The oil seal which is fitted to the fork lower leg is retained by a circlip. **Do not** remove the oil seal unless it is to be renewed because the act of removal will almost certainly damage the fine sealing lip.

5 Front forks: examination and renovation

1   The parts most likely to wear are the oil seals and the bearing surfaces of the lower fork leg and the upper tube. As bushes are not fitted, wear in the fork lower leg can only be remedied by renewal of the complete fork lower leg.
2   Split or perished gaiters or dust covers must be attended to immediately otherwise the ingress of road grit will accelerate wear of the oil seal and upper tube.
3   After an extended period of service the fork springs may take a permanent set. If the spring lengths are suspect, they should be compared with a new set. It is wise to fit new components if the overall length has decreased. Always fit new springs as a pair, NEVER separately.

4 Check the outer surface of the stanchion for scratches or roughness. It is only too easy to damage the oil seal during reassembly, if these high spots are not eased down. The stanchions are unlikely to bend unless the machine is damaged in an accident. Any significant bend will be detected by eye, but if there is any doubt about straightness, roll the stanchion tube on a flat surface. If the stanchions are bent, they must be renewed. Unless specialist repair equipment is available, it is rarely practicable to straighten bent stanchions, and in any event the safety of repaired components of this nature is always suspect.

5 Wear of the damper rod piston ring will reduce damping efficiency and promote deterioration of road holding and handling. The ring may be renewed separately, if required.

6 Steering head bearings: examination and renovation

1 Clean and examine the cups and cones of the steering head bearings. They should have a polished appearance and show no signs of indentation. Renew the set if necessary.

2 Clean and examine the ball bearings which should also be polished and show no signs of surface cracks or blemishes. If any require replacement the whole set must be renewed.

3 All the balls are $\frac{1}{4}$ in diameter (do not mix metric and English sizes as they are slightly different). Nineteen balls are fitted into both top and bottom bearing races. This arrangement will leave a gap but an extra ball must not be fitted otherwise the balls will press against each other, accentuating wear and making the steering stiff.

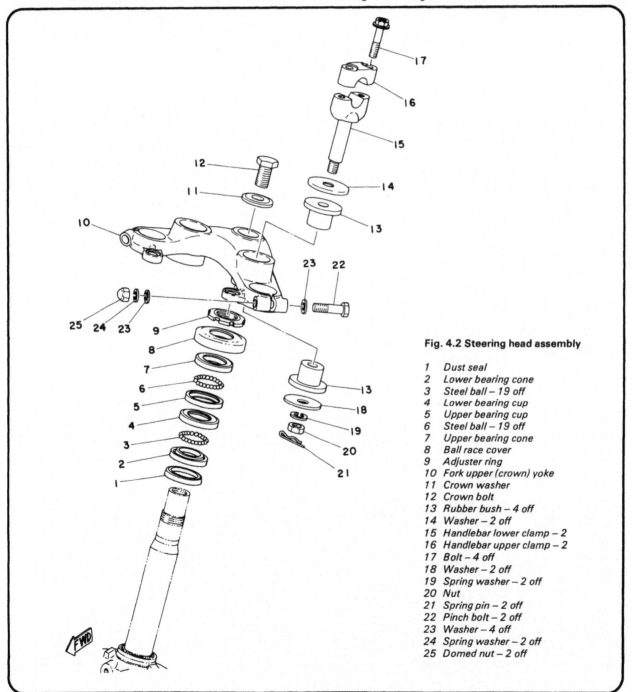

Fig. 4.2 Steering head assembly

1 Dust seal
2 Lower bearing cone
3 Steel ball – 19 off
4 Lower bearing cup
5 Upper bearing cup
6 Steel ball – 19 off
7 Upper bearing cone
8 Ball race cover
9 Adjuster ring
10 Fork upper (crown) yoke
11 Crown washer
12 Crown bolt
13 Rubber bush – 4 off
14 Washer – 2 off
15 Handlebar lower clamp – 2
16 Handlebar upper clamp – 2
17 Bolt – 4 off
18 Washer – 2 off
19 Spring washer – 2 off
20 Nut
21 Spring pin – 2 off
22 Pinch bolt – 2 off
23 Washer – 4 off
24 Spring washer – 2 off
25 Domed nut – 2 off

4.1 Drain plug in fork lower leg

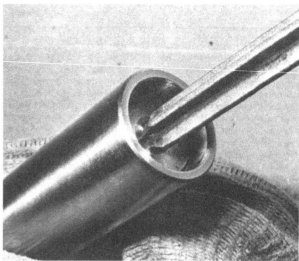

4.2a Depress the top plug to allow circlip removal and ...

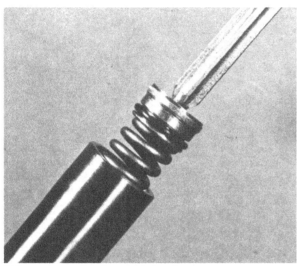

4.2b ... remove the plug and fork spring

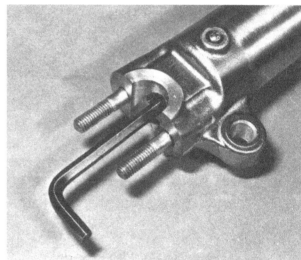

4.3a Unscrew the socket bolt from the lower leg

4.3b Prise off the dust excluder and withdraw ...

4.3c ... the stanchion from the lower leg

4.3d Allow the damper rod to slide out of the stanchion

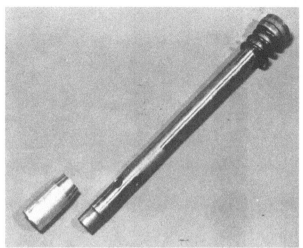

4.3e Damper assembly – general view

5   Overtight head races are equally undesirable. It is possible to unwittingly apply a loading of several tons on the head races when they have been overtightened, even though the handlebars appear to turn quite freely. Overtight bearings will make the machine roll at low speeds and give generally imprecise handling with a tendency to weave. Adjustment is correct if there is no perceptible play in the bearings and the handlebars will swing to full lock in either direction, when the machine is on the centre stand with the front wheel clear of the ground. Only a slight tap should cause the handlebars to swing.

8   Frame: examination and repair

1   The frame is unlikely to require attention unless an accident has caused damage. In this case the frame should be taken to a specialist for repair.
2   It is wise to check the frame occasionally for cracks especially around the welded sections and those subjected to vibration and stress eg; footrest mounting lugs. Rust corrosion can also lead to defects and should be eliminated in its early stages.
3   A frame which is out of alignment will cause handling problems. If this is suspect, it is usually necessary to remove all the components so that the bare frame can be checked and possibly realigned by a specialist in this type of work.

4.4 Oil seal is secured by a circlip

7   Front forks replacement

1   Replace the front forks by following in reverse the dismantling procedures described in Section 2, 3 and 4 of this Chapter. Before fully tightening the front wheel spindle clamps and the fork lower yoke pinch bolts, bounce the forks several times to ensure they work correctly and settle down into their original settings. Complete the final tightening from the wheel spindle upwards.
2   Refill each fork leg with 130 cc (4.4/3.7 US/Imp fl oz) of ATF (automatic transmission fluid) or SAE 10W/30 fork oil; before the plug in the top of each fork leg is replaced. Check that the drain plugs have been re-inserted and tightened before the oil is added!
3   If the fork stanchions prove difficult to relocate through the fork yokes, make sure their outer surfaces are clean and polished so that they will slide more easily. It is often advantageous to use a screwdriver blade to open up the clamps, as the tubes are moved upwards into position.
4   Before the machine is used on the road, check the adjustment of the steering head bearings. If they are too slack, judder will occur, especially during braking. There should be no detectable play in the head races when the handlebars are pulled and pushed with the front brake applied hard.

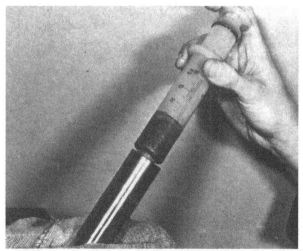

7.2 Refill each fork leg with the correct quantity of damping fluid

1 Frame complete
2 Engine rear mounting plate
3 Engine rear mounting plate
4 Bolt – 4 off
5 Spring washer – 4 off
6 Cylinder head steady plate
7 Cylinder head steady plate
8 Damper – 2 off
9 Bolt – 2 off
10 Nut – 2 off
11 Spring washer – 2 off
12 Bolt – 2 off
13 Bolt – 2 off
14 Bolt
15 Nut – 4 off
16 Nut
17 Spring washer – 4 off
18 Spring washer
19 Steering lock assembly
20 Conical spring
21 Lock cover
22 Wave washer
23 Rivet
24 Cable strap
25 Cable strap
26 Damper
27 Cable strap

7.4 Adjust the steering head bearing using a 'C' spanner

Fig. 4.3 Frame assembly

9 Swinging arm rear fork: removal and dismantling

1   The swinging arm bushes consist of two outer bushes pressed into each end of the fork crossmember which bears on a single inner bush which is retained on the swinging arm pivot spindle. When wear develops in the swinging arm bushes, necessitating renewal, the repair is quite straightforward.

2   To remove the swinging arm, first place the machine on its centre stand with the weight off the rear wheel and remove the wheel as described in Chapter 5 section 3.

3   On disc brake models, detach the two clamps which hold the brake hose to the swinging arm fork. The clamps incorporate a locking tab which secures the heads of the bolts. Bend down the tabs before attempting to loosen each bolt. Remove the pivot bolt which secures the brake caliper unit to the mounting frame. Lift the caliper from position and move it forwards, so that it may be suspended from the frame. If this method of removal is used, there is no necessity to disconnect the caliper from the hydraulic hose.

4   Remove the chainguard which is retained by a single bolt at the rear, and detach both suspension units at their lower mountings.

5   Bend down the ear of the lockwasher which secures the swinging arm pivot bolt nut. Remove the nut and lockwasher. The pivot bolt should be drifted out carefully taking care not to damage the threads. Support the swinging arm fork and withdraw it from the rear of the machine.

10 Swinging arm rear fork: renovation and assembly

1   Remove the dust caps from both ends of the swinging arm fork crossmember and note the number of shims fitted at each side. The shims must be replaced in the same positions on reassembly. Push out the long inner bush.

2   After cleaning the inner bush and the outer bushes thoroughly, insert the inner bush again and check for play. If play is perceptible, the outer bushes must be renewed as a pair. The outer bushes may be drifted from position, using a long drift inserted from the opposite side of the crossmember. The bushes are made of a brittle material which will probably fracture during removal. For this reason they should not be disturbed unless they are to be renewed.

3   Reassemble the swinging arm fork by reversing the dismantling procedure. Grease the pivot shaft and bearings liberally before reassembly as no facility is provided for greasing after assembly.

4   Worn swinging arm pivot bearings will give imprecise handling with a tendency for the rear end of the machine to twitch or hop particularly during the transition from 'power-on' to 'power-off', or vice versa. Play can be detected by placing the machine on its centre stand and with the rear wheel clear of the ground, pulling and pushing on the fork ends in a horizontal direction. Any play will be greatly magnified by the leverage effect.

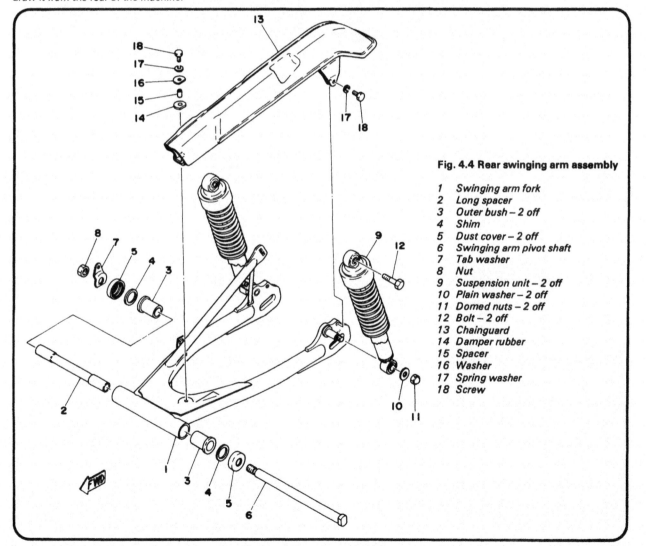

Fig. 4.4 Rear swinging arm assembly

1   Swinging arm fork
2   Long spacer
3   Outer bush – 2 off
4   Shim
5   Dust cover – 2 off
6   Swinging arm pivot shaft
7   Tab washer
8   Nut
9   Suspension unit – 2 off
10 Plain washer – 2 off
11 Domed nuts – 2 off
12 Bolt – 2 off
13 Chainguard
14 Damper rubber
15 Spacer
16 Washer
17 Spring washer
18 Screw

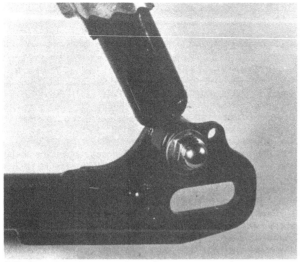

9.5a Detach both suspension units at rear mounting

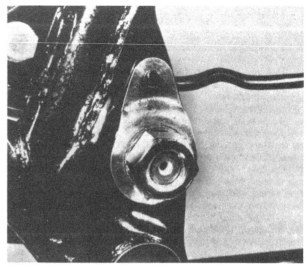

9.5b Bend down tab washer and loosen nut then ...

9.5c ... drive out the pivot shaft

9.5d Lift swinging arm fork away to rear of machine

10.1a Pull off the dust caps and ...

10.1b ... note shims (if any) and sealing ring

10.1c Displace the long bush

10.2 Outer bushes are very brittle and may break on removal

11 Rear suspension units: examination

1   The rear suspension units fitted to the XS250, 360 and 400 models are of the normal hydraulically damped type, adjustable to give five different spring settings. A 'C' spanner should be used to turn the lower spring seat and so alter its position on the adjustment projection. When the spring seat is turned so that the effective length of the spring is shortened, the suspension will become heavier.
2   If a suspension unit leaks, or if the damping efficiency is reduced in any other way the two units must be replaced as a pair. For precise roadholding it is imperative that both units react to movement in the same way. It follows that the units must always be set at the same spring loading.

12 Centre stand: examination

1   The centre stand is attached to the machine by two bolts on the bottom of the frame. It is returned by a centre spring. The bolts and spring should be checked for tightness and tension respectively. A weak spring can cause the centre stand to 'ground' on corners and unseat the rider.

13 Prop stand: examination

1   The prop stand is secured to a plate on the frame with a bolt and nut, and is retracted by a tension spring. Make sure the bolt is tight and the spring not overstretched, otherwise an accident can occur if the stand drops during cornering. For the same reason check that the pivot nut split pin has not been displaced.

14 Footrests: examination and renovation

1   The front footrests are held on the frame by two bolts each, passing into threaded sleeves welded to the frame lugs. A damper rubber placed around each sleeve reduces vibration which might be transmitted to the rider's foot. The rear footrests are an integral part of the two silencer mounting brackets, each of which is secured to the frame on a single stud. The footrests pivot upwards on their mounting brackets and are spring loaded to keep them in their horizontal positions. If an obstacle is struck they will fold upwards, reducing the risk of damage to the rider's foot or to the main frame.

2   If the footrests are damaged in an accident, it is possible to dismantle the assembly into its component parts. Detach each footrest from the frame lugs and separate the folding foot piece from the bracket on which it pivots by withdrawing the split pin and pulling out the pivot shaft. It is preferable to renew the damaged parts, but if necessary, they can be bent straight by clamping them in a vice and heating to a dull red with a blow lamp whilst the appropriate pressure is applied. Do not attempt to straighten the footrests while they are attached to the frame.
3   If heat is applied to the main footrest piece during any straightening operation, it follows that the footrest rubber must be removed temporarily.

15 Rear brake pedal: examination and renovation

1   The rear brake pedal is secured to the shaft on which it pivots by a pinch bolt, and is located by splines. To allow removal of the pedal, the pinchbolt must be removed completely.
2   In the event of damage, the rear brake pedal may be straightened in a manner similar to that prescribed for footrest renovation. Bear in mind that the heat applied may damage the chrome finish.

16 Dualseat: removal and replacement

1   The dualseat is attached to two lugs on the left side of the frame by two clevis pins secured with split pins. If it is necessary to remove the dualseat withdraw the two split pins, take out the clevis pins, and the seat will lift off as a complete unit.

17 Speedometer and tachometer heads: removal and replacement

1   The speedometer and tachometer are both mounted together on a single panel on top of the front forks. They are secured in rubber mounted cases by two nuts with rubber washers over studs mounted on the bottom of the panel.
2   The instruments may be removed individually from their separate cases following the same procedure. Remove the two domed nuts and disconnect the drive cable by unscrewing the knurled ring. Pull the instrument head up and pull out the push fit warning and illumination bulb holders.

3   Unscrew the drive cables and pull out the push fit bulb holders. Check for blown bulbs while they are out.

4   Speedometer and tachometer heads cannot be repaired and if a defect occurs it is best to fit a new instrument. Remember that a speedometer in correct working order is required by law on a machine in the UK and many other countries.

5   If an instrument becomes erratic in operation or fails, suspect first the drive cable, which should be renewed as a complete assembly if found to be damaged.

18 Speedometer and tachometer drives: location and examination

1   The speedometer is driven by a gear that is driven internally from the wheel hub, and is housed inside the brake back plate (drum brake models) or in the case of disc brake models, by a separate gearbox driven from the hub fitted on the front wheel spindle, to the left of the wheel. The drive in both cases should be greased occasionally.

2   The tachometer drive is provided by a gear shaft fitting into the front of the camshaft cover, where it meshes with a scroll gear on the camshaft. As noted in Chapter 1, failure of the gear is rare, as it works in ideal conditions, being lubricated thoroughly and protected from contaminants by full enclosure.

19 Cleaning the machine

1   After removing all surface dirt with warm water and a rag or sponge, use a cleaning compound such as 'Gunk' or 'Jizer' for the oily parts. Apply the cleaner with a brush when the parts are dry so that it has an opportunity to soak into the film of oil or grease. Finish off by washing down liberally, taking care that water does not enter into the carburettors, air cleaners or electrics. If desired, a polish such as Solvol Autosol can be applied to the alloy parts to give them full lustre. Application of a wax polish to the cycle parts and a good chrome cleaner to the chrome parts will also give a good finish. Always wipe the machine down if used in the wet, and make sure the chain is well oiled. There is less chance of water getting into control cables if they are regularly lubricated, which will prevent stiffness of action.

20 Fault diagnosis: frame and forks

Symptom	Cause	Remedy
Machine veers to left or right with hands off handlebars	Wheels out of alignment Forks twisted Frame bent	Check and realign. Strip and repair. Strip and repair or renew.
Machine tends to roll at low speeds	Steering head bearing not adjusted correctly or worn	Check adjustment and renew bearings if necessary.
Machine tends to wander	Worn swinging arm bearings	Check and renew bearings. Check adjustment and renew.
Forks judder when front brake applied	Steering head bearings slack, worn fork components	Renew all worn parts, Check adjustment and renew.
Forks bottom	Short of oil	Replenish with correct viscosity oil.
Fork action stiff	Forklegs out of alignment Bent shafts, or twisted in yokes	Strip and renew or slacken clamp bolts, front wheel spindle and top bolts. Pump forks several times and tighten from bottom upwards.
Machine pitches badly	Defective rear suspension units or ineffective fork damping	Check damping action. Check grade and quantity of oil in front forks.

Chapter 5 Wheels, brakes and tyres

For information relating to later models, see Chapter 7

Contents

Specifications

Tyres	Front	Rear
XS250 ... .,.	3.00 × 18 in	3.50 × 18 in
XS360	3.00 × 18 in	3.50 × 18 in
XS400	3.50 × 18 in	3.50 × 18 in

Brakes		
XS250, XS360 and XS400 ...	Single disc	Single disc
XS360 C	Single disc	Single leading shoe, drum
XS360 2D	Twin leading shoe, drum	single leading shoe, drum

Tyre pressures	Solo	Pillion or continuous high speed
Front	26 psi (1.8 kg/cm²)	28 psi (2.0 kg/cm²)
Rear	28 psi (2.0 kg/cm²)	33 psi (2.3 kg/cm²)

1 General description

All models are fitted with 18 inch diameter wheels carrying a 3·50 in section tyre at the rear, and with the exception of the XS400 model which is fitted with a 3·50 in front tyre, a 3·00 in section tyre at the front.

Wheels are either of the traditional type having steel rims laced to aluminium alloy hubs or are one-piece seven spoke cast units. The wheels fitted depend on the model, date of manufacture and country of original delivery.

All XS250 and 400 models and also the XS360D, are fitted with an hydraulic single disc brake on both wheels. The earlier XS360C model has a single leading shoe drum brake at the rear in place of the disc, and the XS360D, introduced as an economy model, has the drum rear brake and a twin leading shoe front brake.

2 Front wheel: examination and renovation (spoked wheel models)

1 Place the machine on the centre stand so that the front wheel is raised clear of the ground. Spin the wheel and check the rim alignment. Small irregularities can be corrected by tightening the spokes in the affected area, although a certain amount of practice is necessary to prevent over-correction. Any flats in the wheel rim should be evident at the same time. These are more difficult to remove and in most cases it will be necessary to have the wheel rebuilt on a new rim. Apart from the effect on stability, a flat will expose the tyre bead and walls to greater risk of damage.

2 Check for loose or broken spokes. Tapping the spokes is the best guide to tension. A loose spoke will produce a quite different sound and should be tightened by turning the nipple in

an anti-clockwise direction. Always re-check for run-out by spinning the wheel again. If the spokes have to be tightened an excessive amount, it is advisable to remove the tyre and tube by the procedure detailed in Section 21 of this Chapter; this is so that the protruding ends of the spokes can be ground off, to prevent them from chafing the inner tube and causing punctures.

3 Front wheel: examination and renovation (cast alloy wheel models)

1   Carefully check the complete wheel for cracks and chipping, particularly at the spoke roots and the edge of the rim. As a general rule a damaged wheel must be renewed as cracks will cause stress points which may lead to sudden failure under heavy load. Small nicks may be radiused carefully with a fine file and emery paper (No. 600 – No. 1000) to relieve the stress. If there is any doubt as to the condition of a wheel, advice should be sought from a Yamaha repair specialist.
2   Each wheel is covered with a coating of lacquer, to prevent corrosion. If damage occurs to the wheel and the lacquer finish is penetrated, the bared aluminium alloy will soon start to corrode. A whitish grey oxide will form over the damaged area, which in itself is a protective coating. This deposit however, should be removed carefully as soon as possible and a new protective coating of lacquer applied.
3   Check the lateral run out at the rim by spinning the wheel and placing a fixed pointer close to the rim edge. If the maximum run out is greater than 2·0 mm (0·08 in), Yamaha recommend that the wheel be renewed. This is, however, a counsel of perfection; a run out somewhat greater than this can probably be accommodated without noticeable effect on steering. No means is available for straightening a warped wheel without resorting to the expense of having the wheel skimmed on all faces. If warpage was caused by impact during an accident, the safest measure is to renew the wheel complete. Worn wheel bearings may cause rim run out. These should be renewed as described in Section 11 of this Chapter.

4 Front wheel: removal and replacement ·

1   Place the machine on the centre stand and place wooden blocks below the crankcase so that the front wheel is well clear of the ground.
2   Remove the speedometer drive cable from the drive within the front hub by releasing the spring clip (drum brake models) or unscrewing the knurled ring (disc brake models).
3   On drum brake models detach the front brake cable by removing the split pin through the clevis pin passing through the brake operating arm. When the clevis pin is withdrawn, the cable can be detached complete with rubber gaiter and cable adjuster, after the latter has been unscrewed from the brake plate. Removal of the disc brake caliper on disc brake models is not required when detaching the front wheel.
4   The front wheel can now be released by withdrawing the spindle, which passes through the left-hand fork leg and is retained by a castellated nut, and split pin. Note that it will be necessary to slacken the two bolts which hold the clamp around the head of the spindle at the extreme end of the left-hand fork leg. The head of the spindle is drilled, so that a tommy bar can be inserted to aid removal. On drum brake models the wheel will pull clear after the anchorage slot on the brake plate has disengaged from the abutment on the left-hand fork leg.
5   Refit the wheels by reversing the dismantling procedure. Ensure that the speedometer gearbox (disc brake models) and the brake back plate (drum brake models) engage with the fork leg as the wheel is being refitted.

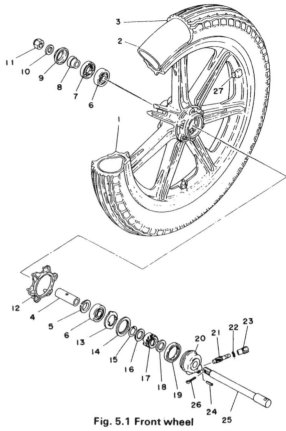

Fig. 5.1 Front wheel

1	Front wheel	15	Circlip
2	Front inner tube	16	Washer
3	Front outer cover	17	Drive gear pinion (17T)
4	Bearing spacer	18	Washer
5	Bearing spacer flange	19	Oil seal
6	Sealed bearing – 2 off	20	Speedometer drive gearbox
7	Oil seal	21	Driven shaft (10T)
8	Spacer	22	Circlip
9	Dust cover	23	Gland union
10	Washer	24	Roll pin
11	Castellated nut	25	Front wheel spindle
12	Housing cover	26	Split pin
13	Speedometer gearbox drive dog	27	Wheel balance weight – 10g or 20g A/R
14	Retainer		

4.3a Remove the nut and spindle to allow ...

4.3b ... the front wheel to be lowered between the forks

4.4a Speedometer gearbox recess must locate with fork projection

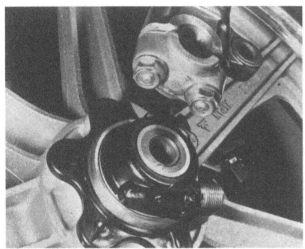

4.4b Wheel spindle clamp must be fitted with arrowmark forwards

5 Front disc brake: pad removal and replacement XS 250, 360 D and 400 models

1   The condition of the brake pads can be determined with the pads in place in the caliper by viewing them through the inspection aperture in the caliper cover. The aperture is closed by a small plastic cap which may be hinged back to give access. Each pad has a red marked groove around the periphery. If it can be seen that one or both pads have worn down to or past the groove, the pads must be renewed as a set.
2   Pad removal can be accomplished without removing the front wheel, as follows. Remove the single bolt which passes through the piston/cylinder casting into the caliper support bracket. It is upon this bolt that the casting slides. From the rear of the unit remove the single crosshead screw by passing the shank of a screwdriver through the wheel spokes. Grasp the piston/cylinder casting and lift it away, leaving the two pads in place on the support bracket, either side of the disc. To prevent the piston being expelled from the cylinder, in the event of the brake lever being applied inadvertently, place a wooden wedge between the piston and outer wall of the casing.
3   Lift each pad away from the disc and out of the support bracket. Note that the outer pad of the XS360C model is fitted

with an anti-chatter shim. This should be detached from the pad.
4   As stated above, the pads must be renewed as a pair if either pad has worn down to the groove. The minimum pad thickness is 1·5 mm (0·06 in).
5   Replace the new pads by reversing the dismantling procedure. On XS360C models, the anti-chatter spring must be fitted to the outer pad so that the stamped arrow mark is facing the direction of normal wheel travel. The outer faces of both pads should be smeared with a thin coating of silicon grease of the type supplied especially for disc brakes. Do not allow any grease to find its way onto the friction surfaces of the pad. It goes without saying that a pad contaminated with grease will not function efficiently. Make sure that the brake pads are correctly located in the caliper and that the front wheel revolves freely when reassembly is complete. Always check the brake action before taking the machine on the road.

6 Front disc brake: removing, renovating and replacing the caliper unit

1   Before the caliper assembly can be removed from the right-hand fork leg, it is first necessary to drain off the hydraulic fluid. Disconnect the brake pipe at the union connection it makes with the caliper unit and allow the fluid to drain into a clean container. It is preferable to keep the front brake lever applied throughout this operation, to prevent the fluid from leaking out of the reservoir. A thick rubber band cut from a section of inner tube will suffice, if it is wrapped tightly around the lever and the handlebars.
2   Note that brake fluid is an extremely efficient paint stripper. Take care to keep it away from any paintwork on the machine or from any clear plastic, such as that sometimes used for instrument glasses.
3   When the fluid has drained off, remove the caliper mounting bolts, then rotate the caliper unit upwards and lift it away from the disc and the machine. Separate the cylinder/piston casting from the support bracket as described for brake pad removal in the preceding Section, and remove the pads.
4   The rubber boot which protects the caliper piston is secured by a circlip. Displace the circlip and pull off the boot. The piston is best expelled from the cylinder by applying a blast of compressed air through the fluid inlet orifice. Take care to catch the piston as it emerges from its bore — if dropped or prised out of position with a screwdriver, it may be damaged irreparably and will have to replaced. Remove the piston seal from the piston.
5   The   parts removed should be cleaned thoroughly, using

only brake fluid as the liquid. Petrol, oil or paraffin will cause the various seals to swell and degrade, and should not be used under any circumstances. When the various parts have been cleaned, they should be stored in polythene bags until reassembly, so that they are kept dust free.

6   Examine the piston for score marks or other imperfections. If there are any imperfections it must be renewed, otherwise air or hydraulic fluid leakage will occur, which will impair braking efficiency. With regard to the various seals, it is advisable to renew them all, irrespective of their appearance. It is a small price to pay against the risk of a sudden and complete front brake failure. It is standard Yamaha practice to renew the seals every two years, even if no braking problems have occurred.

7   Check the condition of the bolt and bush upon which the cylinder casting slides. Renew any component that has worn or corroded.

8   Reassemble under clinically-clean conditions, by reversing the dismantling procedure. The new piston seal should be soaked in clean brake fluid for a few hours to ensure it is supple, before replacement. Lubricate the slide bolt and bush with a graphited or silicon grease and check that the boot has not perished. Reconnect the hydraulic fluid pipe and make sure the union has been tightened fully. Before the brake can be used, the whole system must be bled of air, by following the procedure described in Section 17 of this Chapter

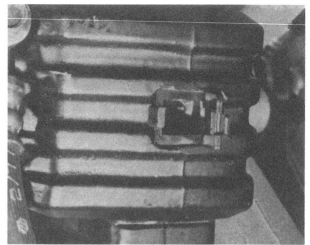
5.1 Disc pad inspection window in caliper

5.2a Remove the caliper slide bolt and ...

5.2b ... the single screw from the rear and ...

5.2c ... pull the caliper casing off the pads

5.3a The outer pad will lift out as will ...

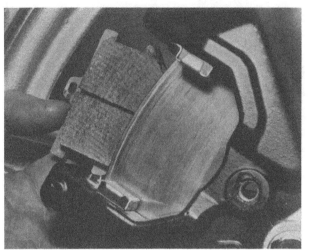

5.3b ... the inner pad

6.6 Seal and boots must be perfect for efficient brake operation

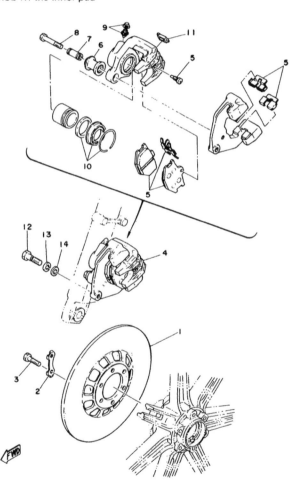

Fig. 5.2 Front disc brake caliper

1	Disc brake	8	Retaining bolt
2	Tab washer – 2 off	9	Bleed nipple
3	Bolt – 6 off	10	Caliper seal kit
4	Caliper assembly complete	11	Inspection cap
5	Caliper pad set	12	Bolt – 2 off
6	Boot	13	Spring washer – 2 off
7	Sleeve	14	Washer – 2 off

7 Master cylinder: examination and renewing seals

1 The master cylinder and hydraulic fluid reservoir take the form of a combined unit mounted on the right-hand side of the handlebars, to which the front brake lever is attached.

2 Before the master cylinder unit can be removed and dismantled, the system must be drained. Place a clean container below the brake caliper unit and attach a plastic tube from the bleed screw of the caliper unit to the container. Lift off the master cylinder cover (cap), gasket and diaphragm, after removing the four countersunk retaining screws. Open the bleed screw one complete turn and drain the system by operating the brake lever until the master cylinder reservoir is empty. Close the bleed screw and remove the tube.

3 Before dismantling the master cylinder, it is essential that a clean working area is available on which the various component parts can be laid out. Use a sheet of white paper, so that none of the smaller parts can be overlooked.

4 Disconnect the stop lamp switch and front brake lever, taking care not to misplace the brake lever return spring. The stop lamp switch is a push fit in the lever stock. The lever pivots on a bolt retained by a single nut. Remove the brake hose by unscrewing the banjo union bolt. Take the master cylinder away from the handlebars by removing the two bolts that clamp it to the handlebars. Take care not to spill any hydraulic fluid on the paintwork or on plastic or rubber components.

5 Withdraw the rubber boot that protects the end of the master cylinder and remove the snap ring that holds the piston assembly in position, using a pair of circlip pliers. The piston assembly can now be drawn out, followed by the return valve, spring cup and return spring.

6 The spring cup can now be separated from the end of the return valve spring and the main cup prised off the piston.

7 Examine the piston and the cylinder cup very carefully. If either is scratched or has the working surface impaired in any other way, it must be renewed without question. Reject the various seals, irrespective of their condition, and fit new ones in their place. It often helps to soften them a little before they are fitted by immersing them in a container of clean brake fluid.

8 When reassembling, follow the dismantling procedure in reverse, but take great care that none of the component parts is scratched or damaged in any way. Use brake fluid as the lubricant whilst reassembling. When assembly is complete, reconnect the brake fluid pipe and tighten the banjo union bolt. Use two new sealing washers at the union so that the banjo bolt does not require overtightening to effect a good seal. Refill the master cylinder with DOT 3 or SAE J1703 brake fluid and bleed the system of air by following the procedure described in Section 17 of this Chapter.

96

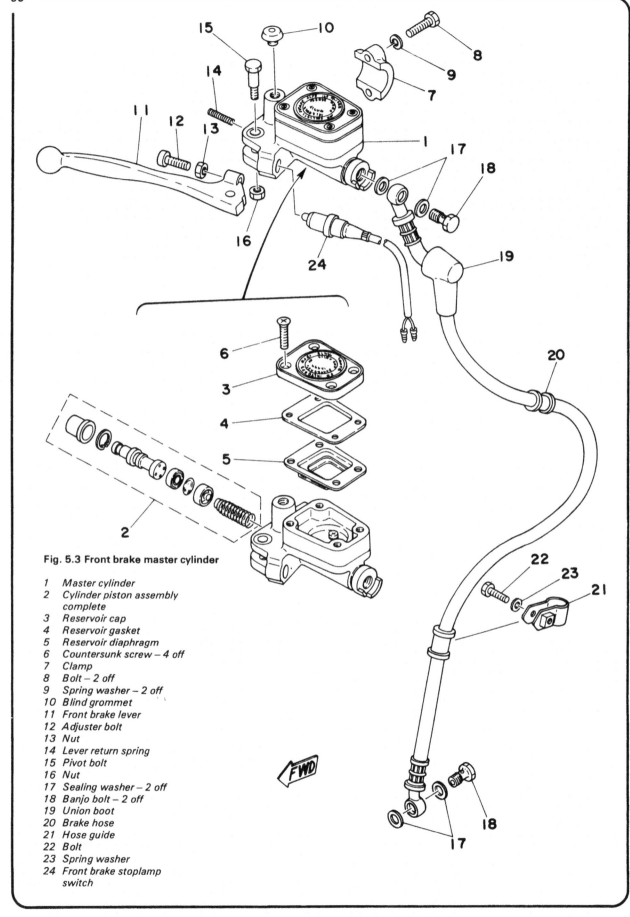

Fig. 5.3 Front brake master cylinder

1   Master cylinder
2   Cylinder piston assembly
    complete
3   Reservoir cap
4   Reservoir gasket
5   Reservoir diaphragm
6   Countersunk screw – 4 off
7   Clamp
8   Bolt – 2 off
9   Spring washer – 2 off
10 Blind grommet
11 Front brake lever
12 Adjuster bolt
13 Nut
14 Lever return spring
15 Pivot bolt
16 Nut
17 Sealing washer – 2 off
18 Banjo bolt – 2 off
19 Union boot
20 Brake hose
21 Hose guide
22 Bolt
23 Spring washer
24 Front brake stoplamp
    switch

8 Removing and replacing the brake disc

1   It is unlikely that the disc will require attention until a considerable mileage has been covered, unless premature scoring of the disc has taken place thereby reducing braking efficiency. To remove the disc, first detach the front wheel as described in Section 4 of this Chapter. The disc is bolted to the front wheel on the right-hand side by six bolts, which are secured in pairs by a common tab washer. Bend back the tab washers and remove the bolts, to free the disc.

2   The brake disc can be checked for wear and for warpage whilst the front wheel is still in the machine. Using a micrometer, measure the thickness of the disc at the point of greatest wear. If the measurement is much less than the recommended service limit of 4·5 mm (0·177 in), the disc should be renewed. Check the warpage of the disc by setting up a suitable pointer close to the outer periphery of the disc and spinning the front wheel slowly. If the total warpage is more than 0·15 mm (0·006 in), the disc should be renewed. A warped disc, apart from reducing the braking efficiency, is likely to cause juddering during braking and will also cause the brake to bind when it is not in use.

8.1 The brake disc is retained by six bolts

9 Front drum brake: examination and renovation

1   With the front wheel removed, as described Section 4 of this Chapter, the twin leading shoe brake mechanism and backplate can be pulled free from the drum.

2   Examine the drum surface for signs of scoring or oil contamination. Both of these conditions will impair braking efficiency. Remove all traces of dust, preferably using a brass wire brush, taking care not to inhale any of it, as it is of an asbestos nature, and consequently toxic. Remove oil or grease deposits, using a petrol soaked rag.

3   If deep scoring is evident, due to the linings having worn through to the shoe at some time, the drum must be skimmed on a lathe, or renewed. Whilst there are firms who will undertake to skim a drum whilst fitted to the wheel, it should be borne in mind that excessive skimming will change the radius of the drum in relation to the brake shoes, therefore reducing the friction area until extensive bedding in has taken place. Also full adjustment of the shoes may not be possible. If in doubt about this point, the advice of one of the specialist engineering firms who undertake this work should be sought.

4   If fork oil or grease from the wheel bearings has badly contaminated the linings, they should be renewed. There is no satisfactory way of degreasing the lining material, which in any case is relatively cheap to replace. It is a false economy to try to cut corners with brake components; the whole safety of both machine and rider being dependent on their condition.

5   The linings are bonded to the shoes, and the shoe must be renewed complete with the new linings. This is accomplished by folding the shoes together until the spring tension is relaxed, and then lifting the shoes and springs off the brake plate. Fitting new shoes is a direct reversal of the above procedure.

6   Before refitting existing shoes, roughen the lining surface sufficiently to break the glaze which will have formed in use.

7   Examine the linkage which runs from the break actuating lever to operate the second fulcrum. This may be removed after slackening the pinch bolts which retain it to the splines. The linkage may be further dismantled, if desired, by removing the circlips and clevises which retain the connecting rod to the actuating levers.

8   Push out the fulcrums from the brake plate. If there is corrosion of the fulcrum face or in its bore, this should be removed using wet or dry paper. Grease both fulcrums before installation. With the shoes, fulcrums and linkage in position on the brake plate, adjust the linkage by means of the turnbuckle and locknut on the connecting rod, so that the fulcrum faces are parallel to each other. This is an important point as both shoes should commence to act on the drum surface at the same point. A maladjusted linkage can result in impaired braking efficiency, whilst a correctly set up twin leading shoe drum brake is both powerful and sensitive in use, giving most of the benefits of a disc brake without the risk of impaired wet weather performance. After reinstallation of the wheel, the front brake should be adjusted finally, as described in the following Section.

10 Adjusting the twin leading shoe front brake

1   If the adjustment of the front brake cable is correct, there should be a clearance of 5 – 8 mm (0·2 – 0·3 in) measured between the brake lever and the lever stock when the brake has just commenced operation. Adjustment may be made either at the handlebar lever cable adjuster or at the adjuster on the brake back plate.

2   The only time the operating rod connecting the two operating levers requires adjustment is when the original setting has been disturbed, or if uneven wear of the brake shoes had led to reduced braking efficiency. It is imperative that the leading edge of each brake shoe contacts the brake drum at the same time, for maximum braking efficiency.

3   Check by detaching the clevis pin from the eye of one end of the operating rod so that the brake operating arms can be applied independently. Operate each arm separately and note when the brake shoe first makes contact with the brake drum surface. Make a mark to show the exact position of each operating arm when the initial contact is made. Replace the clevis pin and check that the marks coincide when the brake is applied in similar fashion. If they do not, withdraw the clevis pin and use the rod adjuster to extend or reduce the length of the operating rod until the marks correspond exactly. Replace the clevis pin and do not omit the 'E' clip over the end which retains the clevis pin in position. Recheck the brake lever adjustment before taking the machine on the road. As a rough guide the two brake operating arms should be parallel with one another, when adjustment is correct.

4   Check that the brake pulls off correctly when the handlebar lever is released. Sluggish action can be due to a poorly lubricated cable or one with a frayed inner.

11 Front wheel bearings: examination and replacement

All models

1   Place the machine on the centre stand and remove the front wheel as described in Section 4. On disc brake machines

remove the speedometer gearbox, drive gear and oil seal on the right-hand side of the hub and remove the spacer, oil seal and bearing retainer circlip from the left-hand side. To give access to the bearings or drum brake machines, first remove the brake plate on the left-hand side of the hub and the dust excluder and oil seal from the right-hand side of the hub.

2   The wheel bearings can now be tapped out from each side with the use of a suitable long drift. Careful and even tapping will prevent the bearing 'tying' and damage to the races.

3   Remove all the old grease from the hub and bearings, giving the latter a final wash in petrol. When the bearings are dirty, lubricate them sparingly with a very light oil. Check the bearings for play and roughness when they are spun by hand. All used bearings will emit a small amount of noise when spun but they should not chatter or sound rough. If there is any doubt about the conditions of the bearings they should be renewed.

4   Before replacing the bearings pack them with high melting point grease. Do not overfill the hub centre with grease as it will expand when hot and may find its way past the oil seals. The hub space should be about 2/3 full of grease. Drift the bearings in, using a drift on the outside ring of the bearing. Do not drift the centre ring of the bearing or damage will be incurred. Replace the oil seals carefully, drifting them into place with a thick walled tube of approximately the same dimension as the oil seal. A large socket spanner is ideal.

11.1a Remove spacers before drifting out bearings

11.1b Oil seals may be prised out or drifted out with bearing

11.4a Do not omit bearing spacer on reassembly

11.4b Drive in bearings using suitable tubular drift

12 Rear wheel: examination and renovation

1   When inspecting the rear wheel follow the procedure given for front wheel inspection in either Section 2 or 3 of this Chapter, depending upon whether the wheel is of the spoked or cast alloy type.

2   The procedure for removal, examination and lubrication of the rear wheel bearings is materially the same as that used when attending to the front wheel bearings. Follow the procedure given in the preceding Section.

13 Rear wheel: removal and replacement

1   Place the machine on the centre stand so that the rear wheel is clear of the ground. Separate the rear chain at the master link, and then replace the link on one end of the chain to avoid loss.

2   Remove the two bolts which secure the chainguard and lift the guard away from the machine towards the rear.

XS 360 drum brake models

3   Detach the brake torque arm from the brake back plate by removing the special bolt, held by a nut and split pin. Unscrew the brake adjuster nut from the brake operating rod and depress the brake pedal, so that the rod leaves the operating arm. Push out the trunnion from the arm and replace it on the rod, together with the adjuster nut. This will prevent loss of these two components and the brake rod spring.

Disc brake models

4   When removing the rear wheel on disc brake models, the rear suspension should be compressed a number of inches, to allow sufficient clearance between the brake caliper and the ground for the disc to be able to leave the caliper. A short length of steel cable with a hook at either end is supplied in the tool kit. After compressing the rear dampers this may be secured on two

lugs provided and so hold the suspension in the compressed position.

All models

5   Remove the split pin from the end of the wheel spindle and unscrew the nut. To free the wheel, withdraw the wheel spindle, if necessary passing a tommy bar through the hole in the spindle head. As the spindle is withdrawn, the two chain adjuster units will fall free, together with the wheel spacer on the right-hand side of the spindle.
6   Lower the wheel and manoeuvre it from position between the swinging arm fork.
7   Refit the rear wheel by reversing the dismantling procedure. On drum brake models it is essential that the torque arm bolt is tightened fully and the split pin refitted. If the arm falls free in service, the rear brake will lock-up. The results of this are obvious.

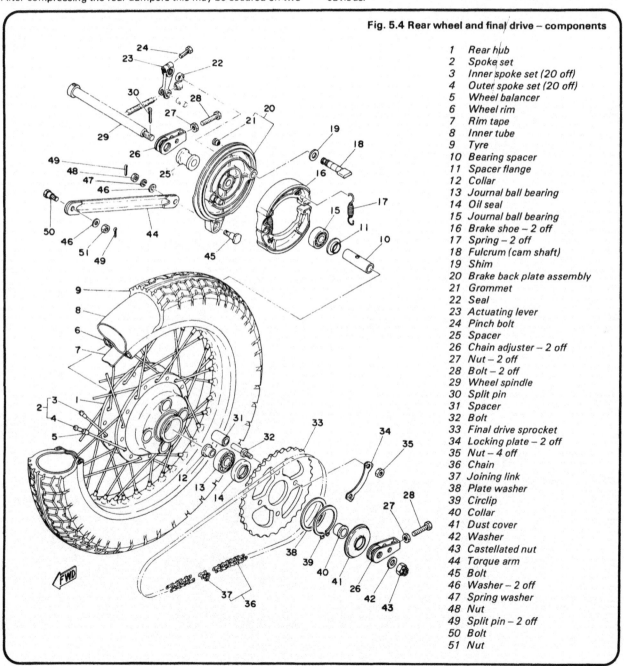

Fig. 5.4 Rear wheel and final drive – components

1   Rear hub
2   Spoke set
3   Inner spoke set (20 off)
4   Outer spoke set (20 off)
5   Wheel balancer
6   Wheel rim
7   Rim tape
8   Inner tube
9   Tyre
10   Bearing spacer
11   Spacer flange
12   Collar
13   Journal ball bearing
14   Oil seal
15   Journal ball bearing
16   Brake shoe – 2 off
17   Spring – 2 off
18   Fulcrum (cam shaft)
19   Shim
20   Brake back plate assembly
21   Grommet
22   Seal
23   Actuating lever
24   Pinch bolt
25   Spacer
26   Chain adjuster – 2 off
27   Nut – 2 off
28   Bolt – 2 off
29   Wheel spindle
30   Split pin
31   Spacer
32   Bolt
33   Final drive sprocket
34   Locking plate – 2 off
35   Nut – 4 off
36   Chain
37   Joining link
38   Plate washer
39   Circlip
40   Collar
41   Dust cover
42   Washer
43   Castellated nut
44   Torque arm
45   Bolt
46   Washer – 2 off
47   Spring washer
48   Nut
49   Split pin – 2 off
50   Bolt
51   Nut

13.5a Remove the split pin and nut to allow ...

13.5b ... the rear wheel spindle to be withdrawn

13.6 Tilt rear wheel and remove from the fork

14 Rear brake: removal, examination and replacement

Disc brake models

1   The rear disc brake caliper and disc are similar components to those used on the front brake of the machine, and for this reason both maintenance and overhaul are fundamentally the same. Refer to the relevant section covering the front brake when carrying out any work.

XS 360 drum rear brake models

The rear drum brake fitted to the XS360C and XS360 – 2D is of the conventional single leading shoe type. When carrying out overhaul or inspection refer to Section 9 which covers the twin leading shoe drum brake unit fitted to the front wheel of XS 360–2D models. Bear in mind that only one fulcrum pin (camshaft) and operating arm is used and hence no link rod is employed.

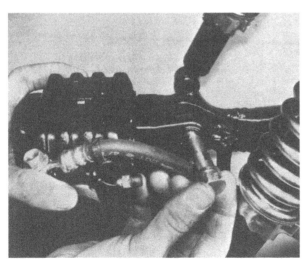

14.1a Rear disc caliper bolt has flat milled for location

14.1b *ALWAYS* renew the caliper bolt split pin

15 Rear brake master cylinder: removal, examination and renewing seals

1   The rear brake master cylinder is attached to the right-hand rear frame downtube and is operated from a foot pedal via a push rod connected to the pedal by a clevis fork and pin. The master cylinder reservoir is integral with the cylinder and has a cap held by three countersunk screws.

2   Disconnect the brake hose at the caliper unit by removing the banjo bolt. Drain the fluid by applying the rear brake until no more fluid emerges. Do not allow fluid to escape onto any cycle parts or rubber components; it is a first class solvent. Detach the hose at the master cylinder where it is held by a banjo bolt.

3   The master cylinder is retained on the frame lug by two bolts. After removal of the bolts, the cylinder unit may be lifted upwards so that the operating pushrod leaves the cylinder.

4   Examination and dismantling of the master cylinder may be carried out in a manner similar to that given for the front brake master cylinder in Section 7.

5   After reassembly and replacement of the rear brake master cylinder components which may be made by reversing the dismantling procedure – bleed the rear brake system of air by referring to Section 17 of this Chapter.

16 Rear brake pedal height: adjustment

1   The pivot shaft upon which the rear brake pedal is mounted is splined to allow adjustment of the pedal height to suit individual requirements.

2   To adjust the height, loosen and remove the pinch bolt which passes into the rear of the pedal. Draw the pedal off the splines and refit it at the required angle. Ideally the pedal should be fitted, so that it is positioned just below the rider's right foot, when the rider is seated normally. In this way the foot does not have to be lifted before the brake can be applied.

3   The upper limit of travel of the brake pedal may be adjusted by means of the bolt and locknut fitted to the pedal pivot mounting bracket. On drum brake models the rear brake adjustment should be checked after making pedal height adjustment by means of the stop bolt.

4   On disc brake models it will be seen that adjustment of pedal height by this method will impart a small degree of movement to the master cylinder piston and so apply the brake slightly. To rectify this, slacken the locknut on the master cylinder pushrod and screw the rod into the clevis fork until free play can be felt between the end of the rod and the master cylinder piston. Unscrew the rod until it contacts the piston lightly and

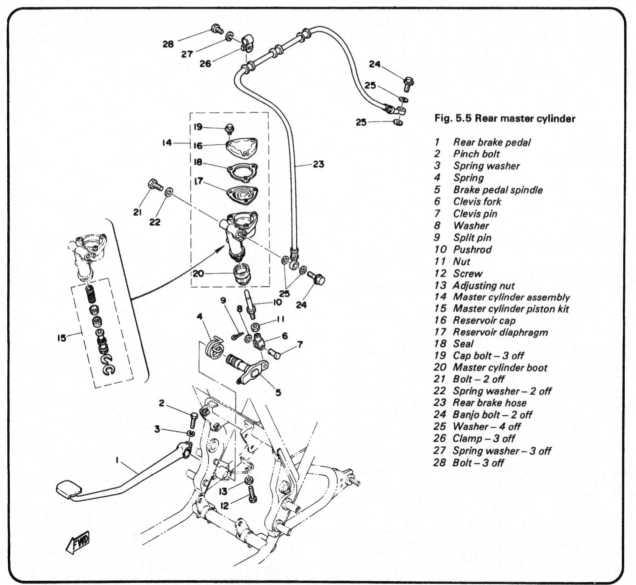

Fig. 5.5 Rear master cylinder

1   Rear brake pedal
2   Pinch bolt
3   Spring washer
4   Spring
5   Brake pedal spindle
6   Clevis fork
7   Clevis pin
8   Washer
9   Split pin
10 Pushrod
11 Nut
12 Screw
13 Adjusting nut
14 Master cylinder assembly
15 Master cylinder piston kit
16 Reservoir cap
17 Reservoir diaphragm
18 Seal
19 Cap bolt – 3 off
20 Master cylinder boot
21 Bolt – 2 off
22 Spring washer – 2 off
23 Rear brake hose
24 Banjo bolt – 2 off
25 Washer – 4 off
26 Clamp – 3 off
27 Spring washer – 3 off
28 Bolt – 3 off

then turn it back 1 – 1½ turns to give the correct free play. Without allowing the pushrod to turn, tighten the locknut. A tiny pin hole in the pushrod is incorporated to show when the pushrod is screwed out of the clevis fork further than is safe. If the pin hole shows above the locknut there is some danger of the rod flexing at the threads. Re-adjust to correct this.

5   Adjustment of the pedal may necessitate readjustment of the rear stop lamp switch.

17 Disc brake: bleeding the hydraulic system

1   Removal of all the air from the hydraulic system is essential to the efficiency of the braking system. Air can enter the system due to leaks or when any part of the system has been dismantled for repair or overhaul. Topping the system up will not suffice, as air pockets will still remain, even small amounts, causing dramatic loss of brake pressure.

2   Check the level in the reservoir, and fill almost to the top. Again, beware of spilling the fluid on to painted or plastic surfaces.

3   Place a clean jar below the brake caliper unit and attach a clear plastic tube from the caliper bleed screw to the container. Place some clean hydraulic fluid in the container so that the pipe is always immersed below the surface of the fluid.

4   Unscrew the bleed screw one complete turn and pump the handlebar lever slowly. As the fluid is ejected from the bleed screw the level in the reservoir will fall. Take care that the level does not drop too low whilst the operation continues, otherwise air will re-enter the system, necessitating a fresh start.

5   Continue the pumping action with the lever until no further air bubbles emerge from the end of the plastic pipe. Hold the brake lever against the handlebars and tighten the caliper bleed screw. Remove the plastic tube **after** the bleed screw is closed.

6   Check the brake action for sponginess, which usually denotes there is still air in the system. If the action is spongy, continue the bleeding operation in the same manner, until all traces of air are removed.

7   When all traces of air have been removed from the system, top up the reservoir and refit the diaphragm and cap or cover, as appropriate. Check the entire system for leaks, and check also that the brake system in general is functioning efficiently before using the machine on the road.

8   Bleeding of the rear brake system should be carried out in a similar manner, using the rear brake pedal in place of the handlebar lever.

9   Brake fluid drained from the system will almost certainly be contaminated, either by foreign matter or more commonly by the absorption of water from the air. All hydraulic fluids are to some degree hygroscopic, that is, they are capable of drawing water from the atmosphere, and thereby degrading their specifications. In view of this, and the relative cheapness of the fluid, old fluid should always be discarded.

18 Rear wheel sprocket: examination and replacement

1   The rear wheel sprocket is retained on the left-hand side of the hub by a large circlip, and is located by four pegs which pass into the cush drive in the hub and are retained by hexagonal nuts.

2   To remove the sprocket, detach the large circlip and the spacing plate which lies below. The sprocket can be removed complete with drive pegs, but it is probable that the pegs are seized in the steel sleeves bonded to the flexible rubbers. If this is the case, knock down the ears of the tab washers and remove the peg securing nuts. The sprocket can then be lifted off with ease.

3   Check the condition of the sprocket teeth. If they are hooked, chipped or badly worn, the sprocket must be renewed. It is considered bad practice to renew one sprocket on its own. The final drive sprockets should always be renewed as a pair and a new chain fitted, otherwise rapid wear will necessitate even earlier renewal on the next occasion.

4   The sprocket may be refitted by reversing the dismantling procedure. It is important that the recesses in the rear of the sprocket are engaged correctly by the milled flats on each cush drive pin.

19 Rear wheel cush drive: examination and renovation

1   The cush drive assembly consists of four tubular rubber bushes located in the hub. The four special pegs retained by nuts on the sprocket locate with these bushes, to give a cushioning effect to the sprocket and drive.

2   To obtain access to the bushes, the sprocket has to be removed by detaching its circlip and pulling it from the wheel hub. Renewal of the bushes is required when there is excessive sprocket movement. As stated in the previous Section the pegs may seize after a considerable length of time. If this occurs, the sprocket complete with pegs should be drawn from the hub using a sprocket puller. A blanking plate or bar, fabricated from mild steel, will have to be made and placed over the bearing. The sprocket puller screw can then bear against the bar.

3   Removal of the flexible bushes is almost impossible without the use of a special expanding extractor. It is recommended that the wheel be returned to a Yamaha dealer who can carry out the work without risk of damage to the wheel.

16.2 A = Pedal height bolt, B = Push rod adjuster nuts

18.2a Displace circlip to remove sprocket and cush drive pins

Tyre changing sequence - tubed tyres

 A Deflate tyre. After pushing tyre beads away from rim flanges push tyre bead into well of rim at point opposite valve. Insert tyre lever adjacent to valve and work bead over edge of rim.

B Use two levers to work bead over edge of rim. Note use of rim protectors

 C Remove inner tube from tyre

 **D** When first bead is clear, remove tyre as shown

 E When fitting, partially inflate inner tube and insert in tyre

 F Work first bead over rim and feed valve through hole in rim. Partially screw on retaining nut to hold valve in place.

 G Check that inner tube is positioned correctly and work second bead over rim using tyre levers. Start at a point opposite valve.

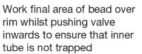

H Work final area of bead over rim whilst pushing valve inwards to ensure that inner tube is not trapped

18.2b To separate sprocket from pins remove nuts

18.2c Note spacers on cush drive pegs

20 Final drive chain: examination and lubrication

1   The final drive chain is fully exposed, with only a light chainguard over the top run. Periodically the tension will need to be adjusted, to compensate for wear. This is accomplished by placing the machine on the centre stand and slackening the wheel nuts on the left-hand side of the rear wheel so that the wheel can be drawn backward by means of the drawbolt adjuster in the fork ends. On drum brake models the torque arm bolt should be slackened after removal of the split pin, to allow the brake back plate to move.
2   The chain is in correct tension if there is approximately 20 mm ($\frac{3}{4}$ inch) slack in the middle of the lower run. Always check when the chain is at its tightest point as a chain rarely wears evenly during service.
3   Always adjust the drawbolts an equal amount in order to preserve wheel alignment. The fork ends are clearly marked with a series of vertical lines above the adjusters, to provide a simple, visual check. If desired, wheel alignment can be checked by running a plank of wood parallel to the machine, so that it touches the side of the rear tyre. If the wheel alignment is correct, the plank will be equidistant from each side of the front wheel tyre, when tested on both sides of the rear wheel. It will not touch the front wheel tyre because this tyre is of smaller cross section. See accompanying diagram.
4   Do not run the chain overtight to compensate for uneven wear. A tight chain will place undue stress on the gearbox and rear wheel bearings, leading to their early failure. It will absorb a surprising amount of power.
5   After a period of running, the chain will require lubrication. Lack of oil will greatly accelerate the rate of wear of both the chain and the sprockets and will lead to harsh transmission. The application of engine oil will act as a temporary expedient, but it is preferable to remove the chain and clean it in a paraffin bath before it is immersed in molten lubricant such as 'Linklyfe' or 'Chainguard'. These lubricants achieve better penetration of the chain links and rollers and are less likely to be thrown off when the chain is in motion.
6   To check whether the chain is due for replacement, lay it lengthwise in a straight line and compress it endwise so that all the play is taken up. Anchor one end and measure the length. Now pull the chain with one end anchored firmly, so that the chain is fully extended by the amount of play in the opposite direction. If there is a difference of more than $\frac{1}{4}$ inch per foot in the two measurements, the chain should be replaced in conjunction with the sprockets. Note that this check should be made AFTER the chain has been washed out, but BEFORE any lubricant is applied, otherwise the lubricant may take up some of the play.

7   When replacing the chain, make sure that the spring link is seated correctly, with the closed end facing the direction of travel.
8   Replacement chains are now available in standard metric sizes from Renold Limited, the British chain manufacturer. When ordering a new chain, always quote the size, the number of chain links and the type of machine to which the chain is to be fitted.

21 Tyres: removal and replacement

1   At some time or other the need will arise to remove and replace the tyres, either as a result of a puncture or because replacements are necessary to offset wear. To the inexperienced, tyre changing represents a formidable task, yet if a few simple rules are observed and the technique learned, the whole operation is surprisingly simple.
2   To remove the tyre from either wheel, first detach the wheel from the machine. Deflate the tyre by removing the valve insert and when it is fully deflated, push the bead from the tyre away from the wheel rim on both sides so that the bead enters the centre well of the rim. Remove the locking cap and push the tyre valve into the tyre itself.
3   Insert a tyre lever close to the valve and lever the edge of the tyre over the outside of the wheel rim. Very little force should be necessary; if resistance is encountered it is probably due to the fact that the tyre beads have not entered the well of the wheel rim all the way round the tyre.
4   Once the tyre has been edged over the wheel rim, it is easy to work around the wheel rim so that the tyre is completely free on one side. At this stage, the inner tube can be removed.
5   Working from the other side of the wheel, ease the other edge of the tyre over the outside of the wheel rim that is furthest away. Continue to work around the rim until the tyre is free completely from the rim.
6   If a puncture has necessitated the removal of the tyre, reinflate the inner tube and immerse in a bowl of water to trace the source of the leak. Mark its position and deflate the tube. Dry the tube and clean the area around the puncture with a petrol soaked rag. When the surface has dried, apply rubber solution and allow this to dry before removing the backing from the patch and applying the patch to the surface.
7   It is best to use a patch of self-vulcanising type, which will form a very permanent repair. Note that it may be necessary to remove a protective covering from the top surfaces of the patch, after it has sealed into position. Inner tubes made from synthetic rubber may require a special type of patch and adhesive, if a satisfactory bond is to be achieved.

8   Before replacing the tyre, check the inside to make sure the agent that caused the puncture is not trapped. Check the outside of the tyre, particularly the tread area, to make sure nothing is trapped that may cause a further puncture.

9   If the inner tube has been patched on a number of past occasions, or if there is a tear or large hole, it is preferable to discard it and fit a replacement. Sudden deflation may cause an accident, particularly if it occurs with the front wheel.

10   To replace the tyre, inflate the inner tube sufficiently for it to assume a circular shape but only just. Then push it into the tyre so that it is enclosed completely. Lay the tyre on the wheel at an angle and insert the valve through the rim tape and the hole in the wheel rim. Attach the locking cap on the first few threads, sufficient to hold the valve captive in its correct location.

11   Starting at the point furthest from the valve, push the tyre bead over the edge of the wheel rim until it is located in the central well. Continue to work around the tyre in this fashion until the whole of one side of the tyre is on the rim. It may be necessary to use a tyre lever during the final stages.

12   Make sure there is no pull on the tyre valve and again commencing with the area furthest from the valve, ease the other bead of the tyre over the edge of the rim. Finish with the area close to the valve, pushing the valve up into the tyre until the locking cap touches the rim. This will ensure the inner tube is not trapped when the last section of the bead is edged over the rim with a tyre lever.

13   Check that the inner tube is not trapped at any point.

Reinflate the inner tube, and check that the tyre is seating correctly around the wheel rim. There should be a thin rib moulded around the wall of the tyre on both sides, which should be equidistant from the wheel rim at all points. If the tyre is unevenly located on the rim, try bouncing the wheel when the tyre is at the recommended pressure. It is probable that one of the beads has not pulled clear of the centre well.

14   Always run the tyres at the recommended pressures and never under or over-inflate. The correct pressures for solo use are given in the Specification Section of this Chapter.

15   Tyre replacement is aided by dusting the side walls, particularly in the vicinity of the beads, with a liberal coating of french chalk. Washing-up liquid can also be used to good effect, but this has the disadvantage of causing the inner surfaces of the wheel rim to rust or corrode.

16   Never replace the inner tube and tyre without the rim tape in position. If this precaution is overlooked there is a good chance of the end of the spoke nipples chafing the inner tube and causing a crop of punctures.

17   Never fit a tyre that has a damaged tread or side walls. Apart from the legal aspects, there is a very great risk of a blow-out, which can have serious consequences on any two-wheeled vehicle.

18   Tyre valves rarely give trouble, but it is always advisable to check whether the valve itself is leaking before removing the tyre. Do not forget to fit the dust cap, which forms an effective second seal.

20.3 Fork ends are marked to aid wheel realignment

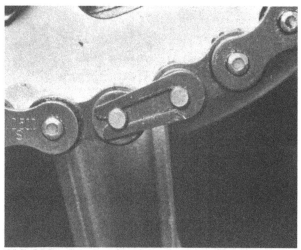

20.7 Spring link closed and must point in direction of chain travel

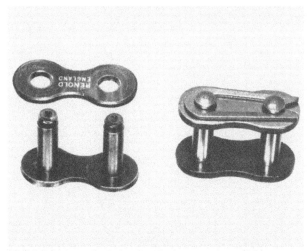

20.8 A British-made chain of equivalent type is available

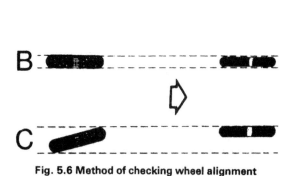

Fig. 5.6 Method of checking wheel alignment

A and C incorrect
B   correct

22 Front wheel balancing

1   The front wheel should be statically balanced, complete with tyre. An out of balance wheel can produce dangerous wobbling at high speed.

2   Some tyres have a balance mark on the sidewall. This must be positioned adjacent to the valve. Even so, the wheel still requires balancing.

3   With the front wheel clear of the ground, spin the wheel several times. Each time, it will probably come to rest in the same position. Balance weights should be attached diametrically opposite the heavy spot, until the wheel will not come to rest in any set position, when spun.

4   Balance weights, which clip round the spokes, are available in 5, 10 or 20 gramme weight. If they are not available, wire solder wrapped round the spokes and secured with insulating tape will make a substitute.

5   It is possible to have a wheel dynamically balanced at some dealers. This requires its removal.

6   There is no need to balance the rear wheel under normal road conditions, although any tyre balance mark should be aligned with the valve.

7   Machines fitted with cast aluminium wheels require special balancing weights which are designed to clip onto the centre rim flange, much in the way that weights are affixed to car wheels. When fitting these weights, take care not to affix any weight near than 40 mm (1·54 in) to the radial centre line of any spoke. Refer to the accompanying diagram.

23 Tyre valve dust caps

1   Tyre valve dust caps are often left off when a tyre has been replaced, despite the fact that they serve an important two-fold function. Firstly they prevent dirt or other foreign matter from entering the valve and causing the valve to stick open when the tyre pump is next applied. Secondly, they form an effective second seal so that in the event of the tyre valve sticking, air will not be lost.

2   Isolated cases of sudden deflation at high speed have been traced to the omission of the dust cap. Centrifugal force has tended to lift the tyre valve off its seating and because the dust cap is missing, there has been no second seal. Racing inner tubes contain provision for this happening because the valve inserts are fitted with stronger springs, but standard inner tubes do not, hence the need for the dust cap.

3   Note that when a dust cap is fitted for the first time, the wheel may have to be rebalanced.

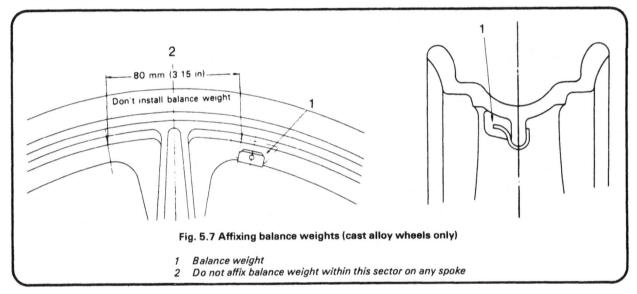

Fig. 5.7 Affixing balance weights (cast alloy wheels only)

1   Balance weight
2   Do not affix balance weight within this sector on any spoke

24 Fault diagnosis: wheels, brakes and tyres

Symptom	Cause	Remedy
Handlebars oscillate at low speeds	Buckle or flat in wheel rim, most probably front wheel	Check rim alignment by spinning wheel. Correct by retensioning spokes or having wheel rebuilt on new rim.
	Tyre not straight on rim	Check tyre alignment.
Machine lacks power and accelerates poorly	Brakes binding	Hot brake drums provide best evidence. Readjust brakes (drum brakes). Caliper slide bolt binding, remove and lubricate bolt and bush.
Brakes grab when applied gently	Ends of brake shoes not chamfered	Chamfer with file.
	Elliptical brake drum (drum brakes)	Lightly skim in lathe (specialist attention needed).
Brake pull-off sluggish	Brake cam binding in housing	Free and grease.
	Weak brake shoe springs	Replace, if brake springs not displaced.
Harsh transmission	Worn or badly adjusted chains	Adjust or replace as necessary.
	Hooked or badly worn sprockets	Replace as a pair, together with chain.

Chapter 6 Electrical System

For information relating to later models, see Chapter 7

Contents

Specifications

	360 2D model	All other models
Battery		
Make	G-S	F-B or Yuasa
Type	12N7-3B-1	12N12-4A-1
Voltage	12V	12V
Capacity	7Ah	12Ah
Earth	Negative	Negative

Alternator	
Type	Excited rotor, fixed coil
Output	14.5V, 13A @ 5,000 rpm
Stator Coil resistance	0.72 ohms ± 10% at 20°C (68°F)
Field Coil resistance	4.04 ohms ± 10% at 20°C (68°F)

Voltage regulator	
Type	Electro-mechanical, three point
Regulating voltage	14.5 ± 0.5V

Starter motor (except XS360 2D)	
Make	Mitsuba
Type	SM223B
Output	0.5Kw
Brush length	11–12.5 mm (0.43–0.49 in)
Wear limit	6.0 mm (0.24 in)
Armature Coil resistance	0.005 ohms ± 10% at 20°C (68°F)
Field Coil resistance	0.011 ohms ± 10% at 20°C (68°F)
Commutator diameter (min)	27 mm (1.06 in)
Mica undercut	0.7 mm (0.03 in)

Bulbs

Headlamp	...	...	...	40/45W (40/30W USA)
Pilot lamp	...	...	...	3W (UK only)
Tail/Stop lamp	...	...	...	8/27W × 2
Flashing indicators	...	...	...	27W × 4
Flasher warning lamp		...	...	3.4W × 2
Instrument light	...	...	...	3.4W × 2
Neutral indicator light		...	...	3.4W
Oil pressure warning light		...	...	3.4W
High beam indicator		...	...	3.4W

All bulbs rated at 12V

1 General description

The Yamaha XS250, 360 and 400 models are all fitted with electrical systems of similar type.

Power is generated by a 12 volt alternator, the fabricated rotor of which is fitted to the extreme left-hand end of the crankshaft. The stator coil is fitted within the alternator cover, together with a field coil which induces a magnetic field in the rotor. The ac current is converted to dc by a silicon rectifier, to allow the battery and electrical ancillaries to be fed. The charging voltage is controlled by an electro-mechanical voltage regulator, which is interconnected with the field coil.

2 Charging system: General

1 As stated above, the electrical charging system consists of an alternator, a rectifier and a voltage control unit. If the charging system output is suspect, the performance of the system as a whole should be checked first. If the output is found lacking, the various components should be checked individually to eliminate the faulty item.

2 The tests described in the following section, all of which relate to the charging system, require the use of a multimeter with resistance functions. It is recommended that unless this equipment is available, and unless some previous experience in its operation has been gained, the machine be returned to a Yamaha Service Agent who will have the necessary equipment and expertise to carry out the work. Irreparable damage to the electrical components may result from injudicious use of the equipment or errors in test wiring.

3 If testing is to be carried out, the following points should be noted:

1) Ensure that the battery is fully charged when making voltage checks.
2) **NEVER** disconnect the leads from the battery whilst the alternator is operating.
3) **NEVER** reverse the polarity of the battery terminals.
4) Take especial care to avoid incorrect connection of wires or short circuiting live wires against earth points.
5) Do not connect the rectifier directly to the battery in an attempt to make a continuity test.
6) Always disconnect the battery before removing electrical components.

Disconnect the two main leads which run from the alternator casing to the block connectors behind the left-hand side panel. Separate the connectors. Select the field winding lead (Green, Green, Orange, Grey) and, using the multimeter set to the resistance function, check the resistance of the field winding across the two green wires. The resistance should be 4.0 ohms ± 15% at 20°C (68°F). If the reading does not conform to that given, the coil is faulty and should be renewed, or a wiring connection has become loose. Select the second main lead from the alternator (three white wires) and test the resistance across each successive pair of wires. Three readings should be taken. Each reading should show a resistance of 0.72 ohms ± 10% at 20°C (68°F). Incorrect readings indicate faulty wiring or damaged coils.

3 When taking resistance readings, it should be noted that variations in ambient temperature will have a proportionate effect on the values that are given.

4 If the resistance value of the coils is found to be correct, the alternator can be considered to be in good condition. Continue the testing of the charging circuit by referring to the following Section.

3.2 Make a visual check of the alternator coil for loose wires

3 Alternator: checking output performance

1 Connect a dc voltmeter, across the two battery terminals. Start the engine and raise the engine speed to 2000 rpm or more. The correct voltage reading is 14.5 ± 0.3V.

2 If a voltage reading above or below the recommended figure is recorded, the alternator winding resistance values should be checked as follows.

4 Silicon rectifier: location and testing

1 The rectifier is located behind the frame left-hand side cover, where it is secured to the battery box by a single screw. The rectifier is a sealed unit, which converts ac current from the alternator to dc current. If the charging system has malfunctioned, but the alternator is found to be in good condition, the rectifier should be checked.

2   Disconnect the rectifier lead at the block connector. Set a multimeter on the resistance function and connect its probes between the red lead and one of the three white leads. Note the reading and then reverse the meter probes. If that particular diode is functioning correctly there should be no continuity shown in one direction with continuity in the other. If either condition exists in both directions then that diode is faulty. Make the same test between the red lead and the other two white leads. Carry out similar tests between the black lead and the three white leads, noting that continuity should only be shown in one direction as described above.
3   If any incorrect readings are given, the rectifier should be renewed.

5   Voltage regulator: operating principle, testing and adjustment

1   The voltage regulator is fitted within the electrical system to control the current from the generator and keep it constant, as engine speed variations would otherwise make the power output rise and fall. By decreasing the field current when the engine speed is high and increasing it when it is low, the generated voltage is maintained at a constant level, regardless of engine speed variations. The voltage regulator has been designed to conform to these operating conditions.
2   If problems with the charging system have developed, but the alternator and rectifier have been tested and found to be in working order, the regulator may be checked using a dc voltmeter with a range of 0-20 volts. The battery should be fully charged for this test.
3   Start the engine and disconnect the wire running from the battery positive (red) terminal to the fuse box. Connect the voltmeter from the wire to a suitable earth point. Increase the engine speed to 2500 rpm and check the reading, which should be 14.5 - 15.0 volts. If the voltage is outside the range given, stop the engine and remove the voltage regulator cover, which is held by two small screws. A small screw bearing against a spring steel plate is provided to allow a limited amount of adjustment to be made to the regulator. Turning the screw clockwise will increase the charging voltage and turning it outwards will reduce the voltage.
4   If the voltage regulator will not respond to adjustment, or functions erratically, the complete unit should be renewed.

5.1 Voltage regulator is mounted below fuse box

6   Battery: examination and maintenance

1   A Furukawa or Yuasa battery of 12 amp hours capacity is fitted as standard to all but the XS360 - 2D model. The XS360-2D has a battery of only 7 amperes per hour capacity because it

is not fitted with a starter motor and therefore does not have to supply additional current.
2   The transparent plastic case of the battery permits the upper and lower levels of the electrolyte to be observed when the battery is fitted from its housing below the dualseat. Maintenance is normally limited to keeping the electrolyte level between the prescribed upper and lower limits and by making sure the vent pipe is not blocked. The lead plates and their separators can be seen through the transparent case, a further guide to the general condition of the battery.
3   Unless acid is spilt, as may occur if the machine falls over, the electrolyte should always be topped up with distilled water, to restore the correct level. If acid is spilt on any of the machine, it should be neutralised with an alkali such as washing soda and washed away with plenty of water, otherwise serious corrosion will occur. Top up with sulphuric acid of the correct specific gravity (1.260 - 1,280) only when spillage has occurred. Check that the vent pipe is well clear of the frame tubes or any of the other cycle parts, for obvious reasons.

7   Battery: charging procedure

1   The correct charging rate for the 12 Ah battery is 1.2 amps, and for the 7 Ah battery is 0.7 amps. A higher charge rate should, if possible, be avoided since it will shorten the working life of the battery.
2   Make sure that the battery charger connections are correct, red to positive and black to negative. It is preferable to remove the battery from the machine whilst it is being charged and to remove the vent plug from each cell. When the battery is re-connected to the machine, the black lead must be connected to the negative terminal and the red lead to positive. This is most important, as the machine has a negative earth system. If the terminals are inadvertently reversed, the electrical system will be damaged permanently.

8   Fuse: location and replacement

1   A bank of fuses is contained within a small plastic box located to the rear of the battery box. The box contains four 10A fuses and two 20A fuses, of which one of each type is spare.
2   Before replacing a fuse that has blown, check that no obvious short circuit has occurred, otherwise the replacement fuse will blow immediately it is inserted. It is always wise to check the electrical circuit thoroughly, to trace the fault and eliminate it.
3   When a fuse blows while the machine is running and no spare is available, a 'get you home' remedy is to remove the blown fuse and wrap it in silver paper before replacing it in the fuseholder. The silver paper will restore the electrical continuity by bridging the broken fuse wire. This expedient should **NEVER** be used if there is evidence of a short circuit or other major electrical fault, otherwise more serious damage will be caused. Replace the 'doctored' fuse at the earliest possible opportunity, to restore full circuit protection.

9   Headlamp: replacing bulbs and adjusting beam height

1   In order to gain access to the headlamp unit the rim must be removed. Remove the two screws from just behind the rim in the 8 o'clock and 4 o'clock positions and pull the rim out at the lower edge. The rim complete with the unit will pull free. Pull the wiring socket off the rear of the reflector unit.
2   On UK models the headlamp bulb is secured in the rear of the reflector unit by a spring loaded bayonet fitting collar or by a plastic screw collar. The bulb in the former type may be released by depressing the spring loaded collar and twisting it to the left. Release the collar and pull out the bulb. On the latter unit

unscrew the collar. In both cases the bulb locates positively in the reflector so that it can be refitted only in one position.

3   The pilot light holder is a push fit in the reflector, the bulb being of the bayonet fitting type. To release, press the bulb inwards and twist to the left.

4   On machines of USA specification no bulb as such is fitted. A single sealed unit comprising the headlamp glass reflector and the twin light filament is utilised. Consequently if either or both filaments fail, the complete unit must be renewed. No provision is made for a pilot bulb.

5   To remove the rim from the sealed beam unit and the unit housing, first detach the long adjuster screw which passes through the rim on the left-hand side. Note the sequence of the spacer washer and spring. Unscrew the two screws at the upper and lower rim brackets and detach the rim. The replacement sealed unit can be fitted by reversing the dismantling procedure.

6   Vertical adjustment of the headlamp can be made by slackening the two mounting bolts. Provision for horizontal adjustment is given only on sealed beam units and is made using the screw in the headlamp rim. The headlamp adjustments should be arranged so that the main beam will not dazzle a person standing at a distance greater than 25 feet from the lamp, whose eye level is not less than 3 feet 6 inches above that plane.

7   To obtain the correct beam height, place the machine on level ground facing a wall 25 feet distant, with the rider seated normally. The height of the beam centre should be equal to the height of the centre of the headlamp from the ground, when the dip switch is in the main beam position. Furthermore, the concentrated area of light should be centrally disposed,. Adjustments in either direction are made by rearranging the angle of the headlamp, as described in the preceding paragraph. Note that a different beam setting will be needed when a pillion passenger is carried. If a pillion passenger is carried regularly,

the passenger should be seated in addition to the rider when the beam setting adjustment is made.

8   The above instructions for beam setting relate to the requirements of the United Kingdom's transport lighting regulations. Other settings may be required in countries other than the UK.

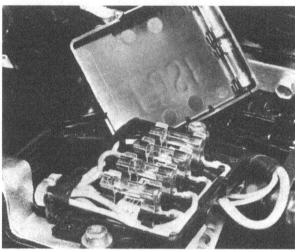

8.1 Spare fuses as contained in the fuse box lid

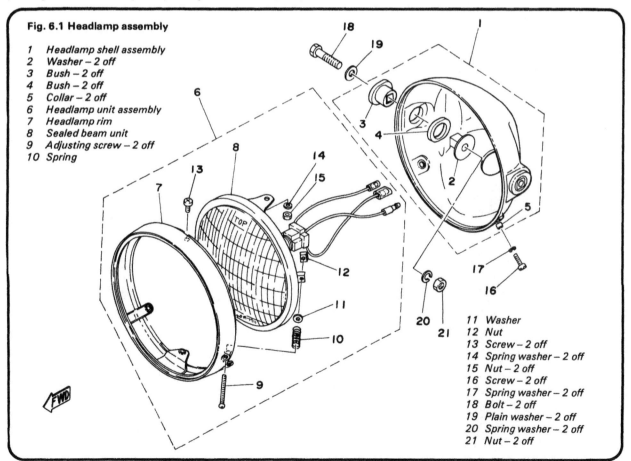

Fig. 6.1 Headlamp assembly

1   Headlamp shell assembly
2   Washer – 2 off
3   Bush – 2 off
4   Bush – 2 off
5   Collar – 2 off
6   Headlamp unit assembly
7   Headlamp rim
8   Sealed beam unit
9   Adjusting screw – 2 off
10   Spring

11   Washer
12   Nut
13   Screw – 2 off
14   Spring washer – 2 off
15   Nut – 2 off
16   Screw – 2 off
17   Spring washer – 2 off
18   Bolt – 2 off
19   Plain washer – 2 off
20   Spring washer – 2 off
21   Nut – 2 off

9.1 Remove screws passing through shell to release reflector unit

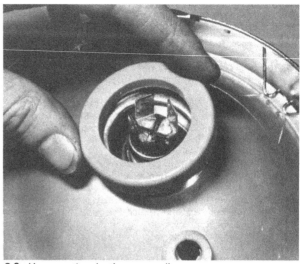

9.2a Unscrew the plastic screw collar

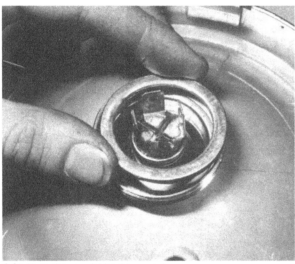

9.2b Note the special backing washers

9.2c Projection on bulb holder locates bulb in correct position

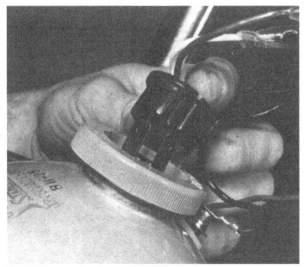

9.2d Wiring socket is a push fit on bulb pins

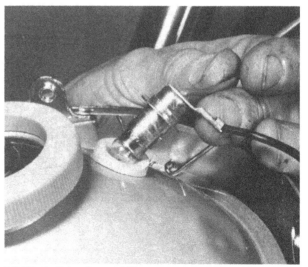

9.3 Pilot bulb is a bayonet fit

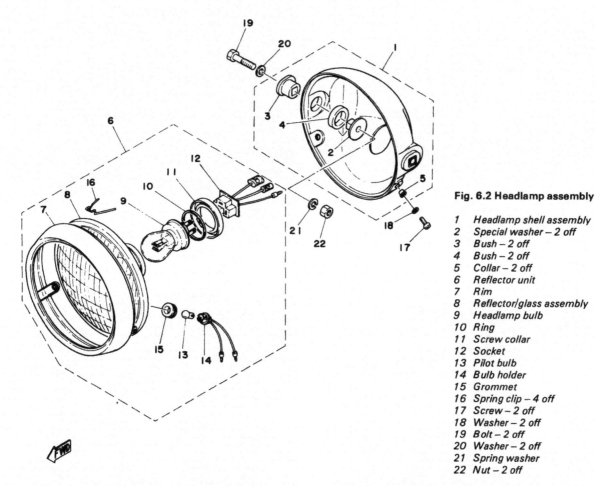

Fig. 6.2 Headlamp assembly

1   *Headlamp shell assembly*
2   *Special washer – 2 off*
3   *Bush – 2 off*
4   *Bush – 2 off*
5   *Collar – 2 off*
6   *Reflector unit*
7   *Rim*
8   *Reflector/glass assembly*
9   *Headlamp bulb*
10 *Ring*
11 *Screw collar*
12 *Socket*
13 *Pilot bulb*
14 *Bulb holder*
15 *Grommet*
16 *Spring clip – 4 off*
17 *Screw – 2 off*
18 *Washer – 2 off*
19 *Bolt – 2 off*
20 *Washer – 2 off*
21 *Spring washer*
22 *Nut – 2 off*

10 Handlebar switches: function and replacement

1   The dipswitch forms part of the left-hand dummy twist grip which contains the horn button and the indicator lamp switch. The right-hand twist grip assembly incorporates the lighting master switch and a three position ignition positive cut-out switch.

2   In the event of failure of any of these switches, the switch assembly must be replaced as a complete unity since it is not practicable to effect a permanent repair.

11 Stop and tail lamp: replacing the bulbs

1   The tail lamp is fitted with two twin filament bulbs of 12 volt, 8/23 rating, to illuminate the rear number plate and rear of the machine, and to give visual warning when the rear brake is applied. To gain access to the bulbs remove the plastic lens cover, which is retained by two long screws. Check that the gasket between the lens cover and the main body of the lamp is in good condition.

2   The bulbs have a bayonet fitting and have staggered pins to prevent the bulb contacts from being reversed.

3   If the tail lamp bulbs keep blowing, suspect either vibration of the rear mudguard or more probably, an intermittent earth connection.

12 Flashing indicator lamps

1   The forward facing indicator lamps are connected to stalks which are attached to the ends of the fork shrouds. The hollow stalks carry the leads to the lens unit. The rear facing lamps are mounted on similar, shorter stalks, at a point immediately to the rear of the dualseat.

2   In each case, access to the bulb is gained by removing the plastic lens cover, which is retained by two screws. Bayonet fitting bulbs of the single filament type are used, each with a 12 volt/27W rating.

13 Flasher unit: location and replacement

1   The flasher relay unit is located under the dualseat, being rubber-mounted on the frame.

2   If the flasher unit is functioning correctly, a series of audible clicks will be heard when the indicator lamps are in action. If the unit malfunctions and all the bulbs are in working order, the usual symptom is one initial flash before the unit goes dead; it will be necessary to replace the unit complete if the fault cannot be attributed to any other cause.

3   In addition to the flasher unit, an electronic flasher cancelling unit is incorporated in the indicator system on all but XS360 - 2D models. The unit automatically turns the flasher light off a certain time after the flasher switch has been operated. The time lapse is dependent on the speed of the machine. If the machine is travelling fast, the unit cancels automatically after a short time. The slower the machine is travelling, the longer the time taken for cancellation. The system may be overriden manually by pushing the indicator switch button inwards.

4   Take great care when handling either unit because they are easily damaged if dropped.

11.1a Stop/tail lamp lens is held by two screws as are ...

11.1b ... the flashing indicator lamp lenses

11.1c Stop/tail lamp has two double filament bulbs

13.1 Flasher unit located forward of battery box

13.3 Self-cancelling flasher unit is located above flasher unit

14 Starter motor: removal, examination and replacement

All except XS 360-2D models

1   An electric starter motor, operated from a small push-button on the right-hand side of the handlebars, provides an alternative and more convenient method of starting the engine, without having to use the kickstart. The starter motor is mounted within a compartment at the rear of the cylinder block, closed by an oblong, chromium plated cover. Current is supplied from the battery via a heavy duty solenoid switch and a cable capable of carrying the very high current demanded by the starter motor on the initial start-up.

2   The starter motor drives a free running clutch immediately behind the generator rotor. The clutch ensures the starter motor drive is disconnected from the primary transmission immediately the engine starts. It operates on the centrifugal principle; spring loaded rollers take up the drive until the centrifugal force of the rotating engine overcomes their resistance and the drive is automatically disconnected.

3   To remove the starter motor from the engine unit, first disconnect the positive lead from the battery, then the starter motor cable from the solenoid switch. Pull the starter cable down through the frame so that it may be detached still connected to the starter motor. The alternator cover must be removed from the left-hand side of the engine to enable remeshing the starter drive sprocket with the splined starter shaft, during reassembly. Remove the two bolts in the chromium plated cover over the starter motor housing and lift the cover away, complete with gasket. The starter motor is secured to the crankcase by two bolts which pass through the right-hand end of the motor casing. When these bolts are withdrawn, the motor can be prised out of position and lifted out of its compartment, with the heavy duty cable still attached.

4   The parts of the starter motor most likely to require attention are the brushes. The end cover is retained by the two long screws which pass through the lugs cast on both end pieces. If the screws are withdrawn, the end cover can be lifted away and the brush gear exposed.

5   Lift up the spring clips which bear on the end of each brush and remove the brushes from their holders. Each brush should have a length of 11.0 - 12.5 mm (0.43 - 0.49 in). The minimum allowable brush length is 6 mm (0.25 in). If the brush is shorter it must be renewed.

6   Before the brushes are replaced, make sure that the commutator is clean. The commutator is the copper segments on which the brushes bear. Clean the commutator with a strip of glass paper. Never use emery cloth or 'wet-or-dry' as the small abrasive fragments may embed themselves in the soft brass of the commutator and cause excessive wear of the brushes. Finish off the commutator with metal polish to give a smooth surface and finally wipe the segments over with a methylated spirits soaked rag to ensure a grease free surface. Check that the mica insulators, which lie between the segments of the commutator are undercut. The standard groove depth is 0.7 mm (0.028 in) but if the average groove depth is less than 0.2 mm (0.008 in) the armature should be renewed or returned to a Yamaha Service Agent for re-cutting.

7   Replace the brushes in their holders and check that they slide quite freely. Make sure the brushes are replaced in their original positions because they will have worn to the profile of the commutator. Replace and tighten the end cover, then replace the starter motor and cable in the housing, tighten down and remake the electrical connection to the solenoid switch. Check that the starter motor functions correctly before replacing the compartment cover and sealing gasket.

15 Starter solenoid switch: function and location

1   The starter motor switch is designed to work on the electromagnetic principle. When the starter motor button is depressed, current from the battery passes through windings in the switch solenoid and generates an electro-magnetic force which causes a set of contact points to close. Immediately the points close, the starter motor is energised and a very heavy current is drawn from the battery.

2   This arrangement is used for at least two reasons. Firstly, the starter motor current is drawn only when the button is depressed and is cut off again when pressure on the button is released. This ensures minimum drainage on the battery. Secondly, if the battery is in a low state of charge, there will not be sufficient current to cause the solenoid contacts to close. In consequence, it is not possible to place an excessive drain on the battery which, in some circumstances, can cause the plates to overheat and shed their coatings. If the starter will not operate, first suspect a discharged battery. This can be checked by trying the horn or switching on the lights. If this check shows the battery to be in good shape, suspect the starter switch which should come into action with a pronounced click. It is located below the dualseat to the right of the battery box and can be identified by the heavy duty starter cable connected to it. It is not possible to effect a satisfactory repair if the switch malfunctions; it must be renewed.

16 Starter motor free running clutch: examination and renovation

1   To gain access to the starter clutch, the alternator cover and the alternator rotor must be removed as described in Chapter 1, Section 9.

2   To check whether the clutch is operating freely and smoothly push the starting sprocket into the clutch (and hence the rotor) and rotate it whilst holding the rotor. If the sprocket is rotated clockwise as viewed from the sprocket side it should lock the clutch. If rotated in an anti-clockwise direction it should be free to run smoothly. If the movement is unsatisfactory, check the condition of the rollers and springs.

3   The rollers, plungers and springs may be prised from position for inspection. If there is any wear or obvious damage to the springs or rollers new replacements will have to be obtained. Replacement of these items requires some dexterity due to spring loading and obstructed access to the spring housings.

14.6a Check condition of commutator segments and ...

14.6b ... inspect brushes for excessive wear

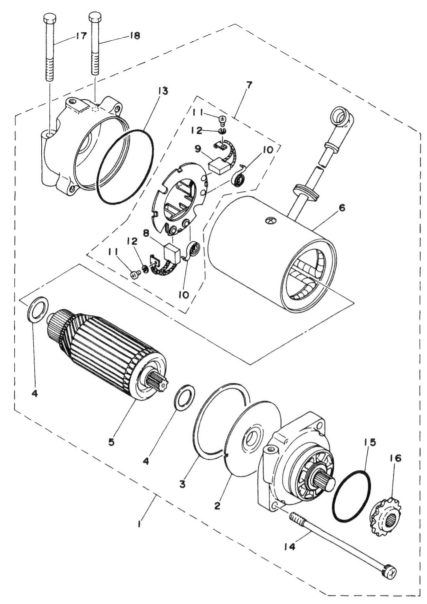

Fig. 6.3 Starter motor

1   Starter motor assembly complete
2   Plate
3   Gasket
4   Washer
5   Armature
6   Yoke
7   Brush holder
8   Positive brush
9   Negative brush
10 Brush spring – 2 off
11 Screw – 2 off
12 Washer – 2 off
13 'O' ring
14 Screw – 2 off
15 'O' ring
16 Drive sprocket
17 Bolt
18 Bolt

15.1 Starter solenoid is mounted in rubber carrier

16.3 Check rollers and springs for damage

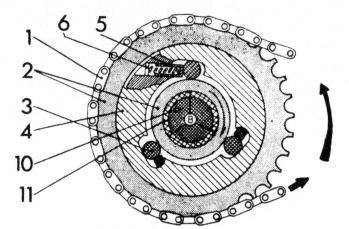

Fig. 6.4 Starter motor free running clutch

1   *Starting chain*
2   *Starting sprocket*
3   *Roller*
4   *Clutch outer*
5   *Roller spring cap*
6   *Roller spring*
10 *Left crankshaft*
11 *Bush*

17 Speedometer and tachometer head: replacement of bulbs

1   The speedometer and tachometer head contain three and two bulbs respectively, all of which are rated 12v 3.4W and are of the bayonet fitting type.
2   The bulbholders are a push fit into the base of the instrument where they are retained by their outer moulded rubber sleeves. Access to the bulbholders can be made after removing the two dome nuts which hold each head in place in its separate outer shell.

18 Warning lamp console: replacement of bulbs

1   In addition to the five bulbs fitted within the instrument heads, a further four warning bulbs are fitted within a warning console to the rear of the instrument assembly. Again, each bulb is rated at 3.4W and is of the bayonet type.
2   Access to the bulbs may be made after removing the console top cover, which is held in place by four screws.

19 Stop lamp switches: location and replacement

1   Two stop lamp switches are fitted to the machine, which work independently of one another, depending on which brake is operated.
2   The front brake switch is fitted to the handlebar lever stock and is a mechanical push-off type, being operated when the lever is moved. The switch is a push fit in the housing boss, and is detached by depressing a small pin in the underside with a piece of wire or a small screwdriver.
3   The rear brake switch is mounted on the frame on the right-hand side, above the rear brake pedal. It can be adjusted by means of a locknut, and should be set so that the light comes on as soon as the pedal is depressed. This is especially important when the rear brake has been readjusted.

20 Horn: location and examination

1   The horn is located on the top frame tube, immediately below the petrol tank. It has no external means of adjustment. If it malfunctions, it must be renewed; it is a statutory requirement that the machine must be fitted with a horn in working order.

21 Wiring: layout and examination

1   The wiring harness is colour-coded and will correspond with the accompanying wiring diagram. Where socket connectors are used, they are designed so that reconnection can be made in the correct position only.
2   Visual inspection will show whether there are any breaks or frayed outer coverings which will give rise to short circuits. Another source of trouble may be the snap connectors and sockets, where the connector has not been pushed fully home in the outer housing.
3   Intermittent short circuits can often be traced to a chafed wire that passes through or is close to a metal component such as a frame member. Avoid tight bends in the lead or situations where a lead can become trapped between casings.

22 Ignition and lighting switch

1   The ignition and lighting switch is combined in one unit, bolted to the top fork yoke. It is operated by a key, which cannot be removed when the ignition is switched on.
2   The number stamped on the key will match the number of the steering head lock and that of the lock in the petrol filler cap. A replacement key can be obtained if the number is quoted; if either of the locks or the ignition switch is changed, additional keys will be required.
3   It is not practicable to repair the ignition switch if it malfunctions. It should be renewed with a new switch and key to suit.

23 Fault diagnosis: electrical system

Symptom	Cause	Remedy
Complete electrical failure	Blown fuse	Check wiring and electrical components for short circuit before fitting new fuse.
	Isolated battery	Check battery connections, also whether connections show signs of corrosion.
Dim lights, horn inoperative	Discharged battery	Recharge battery with battery charger and check whether alternator is giving correct output (electrical specialist).
Constantly 'blowing' bulbs	Vibration, poor earth connection	Check whether bulb holders are secured correctly. Check earth return or connections to frame.

Chapter 7 The 1978 to 1984 UK models and 1977 to 1982 XS400 US models

Contents

Specifications

The following specifications relate to the 1978 to 1984 UK models and the 1977 to 1982 US models. All models are identified by their code numbers; refer to Section 1 for details of machine identification.

Specifications relating to Chapter 1

Engine

Type	Air-cooled twin-cylinder sohc four-stroke
Bore:	
XS400 model	69.0 mm (2.72 in)
XS250 model	55.0 mm (2.17 in)
Stroke	52.4 mm (2.06 in)
Capacity:	
XS400 models	391 cc (23.85 cu in)
XS250 models	248 cc (15.13 cu in)
Compression ratio (UK models):	
XS400 (3N7)	9.4:1
XS250 (3N6)	9.6:1
All other models	9.3:1
Compression ratio (US models):	
XS400 D (2A2)	9.2:1
XS400 2E (2G5)	9.4:1
All other models	9.3:1

Cylinder barrel
Bore size:
XS250 models	55.00 – 55.02 mm (2.1700 – 2.1708 in)
XS400 models	69.00 – 69.02 mm (2.7160 – 2.7173 in)
Bore taper limit	0.05 mm (0.002 in)
Bore ovality limit	0.01 mm (0.0004 in)

Pistons
Piston to bore clearance	0.030 – 0.050 mm (0.0012 – 0.0019 in)
Piston oversizes	+0.25 mm, +0.50 mm, +0.75 mm and +1.00 mm

Gudgeon (piston) pin:
Outside diameter	16.000 – 16.005 mm (0.6300 – 0.6302 in)

Length:
XS250 models	46.4 – 46.7 mm (1.827 – 1.839 in)
XS400 models	58.2 – 58.5 mm (2.288 – 2.300 in)

Piston rings
Thickness:
Top	1.0 mm (0.04 in)
2nd	1.5 mm (0.06 in)
Oil	2.45 mm (0.096 in)

End gap (installed):

Top and second:
XS250 models	0.15 – 0.35 mm (0.006 – 0.014 in)
XS400 models	0.2 – 0.4 mm (0.008 – 0.016 in)
Oil	0.2 – 0.9 mm (0.008 – 0.035 in)

Ring to groove clearance:
Top	0.04 – 0.08 mm (0.0016 – 0.0032 in)
2nd	0.03 – 0.07 mm (0.0012 – 0.0028 in)

Camshaft
Bearing surface ID	23.000 – 23.021 mm (0.9060 – 0.9142 in)
Shaft journal OD	22.967 – 22.980 mm (0.9047 – 0.9052 in)
Clearance	0.020 – 0.054 mm (0.00079 – 0.00213 in)

Cam lobe overall height (service limit):

XS400 SG (3F9) and G (3F8):
Inlet	39.33 mm (1.548 in)
Exhaust	39.37 mm (1.550 in)

All other models:
Inlet	39.38 mm (1.550 in)
Exhaust	39.42 mm (1.552 in)

Base circle diameter (service limit):

XS400 SG (3F9) and G (3F8):
Inlet	32.11 mm (1.264 in)
Exhaust	31.96 mm (1.258 in)

All other models:
Inlet	32.12 mm (1.265 in)
Exhaust	31.97 mm (1.259 in)

Cam lift:

XS400 SG (3F9) and G (3F8):
Inlet	7.53 mm (0.296 in)
Exhaust	7.53 mm (0.296 in)

All other models:
Inlet	7.53 mm (0.296 in)
Exhaust	7.57 mm (0.298 in)

Valve timing
	XS400	XS250
Inlet opens at	30° BTDC	35° BTDC
Inlet closes at	70° ABDC	65° ABDC
Exhaust opens at	70° BBDC	75° BBDC
Exhaust closes at	30° ATDC	25° ATDC

Rocker arms and shafts
XS400 SH (4R4) and H (4R5):
Rocker arm bore	13.000 – 13.018 mm (0.5120 – 0.5127 in)
Rocker shaft diameter	12.985 – 12.991 mm (0.5112 – 0.5115 in)
Shaft to rocker clearance	0.009 – 0.033 mm (0.00035 – 0.00130 in)

All other models:
Rocker arm bore	13.000 – 13.018 mm (0.5120 – 0.5127 in)
Rocker shaft diameter	12.964 – 12.984 mm (0.5104 – 0.5118 in)
Shaft to rocker clearance	0.016 – 0.054 mm (0.00063 – 0.00213 in)

Valves and springs

Valve clearances (cold):
- Inlet .... 0.08 – 0.12 mm (0.0031 – 0.0047 in)
- Exhaust .... 0.16 – 0.20 mm (0.0063 – 0.0078 in)

Valve stem diameter:
- Inlet .... 6.975 – 6.990 mm (0.2746 – 0.2752 in)
- Exhaust .... 6.955 – 6.970 mm (0.2738 – 0.2744 in)

Valve guide diameter:
- Inlet .... 7.000 – 7.012 mm (0.2756 – 0.2761 in)
- Exhaust .... 7.000 – 7.012 mm (0.2756 – 0.2761 in)

Valve stem to guide clearance:
- Inlet .... 0.010 – 0.037 mm (0.0004 – 0.0015 in)
- Exhaust .... 0.030 – 0.057 mm (0.0012 – 0.0022 in)

Valve spring free length:
- Inner .... 39.3 mm (1.547 in)
- Outer .... 42.8 mm (1.685 in)

Crankshaft

Big-end bearing:
- Type .... Plain bearing
- Oil clearance .... 0.021 – 0.045 mm (0.0008 – 0.0018 in)
- Side clearance .... 0.160 – 0.264 mm (0.0063 – 0.0104 in)
Small-end ID .... 16.015 – 16.028 mm (0.6305 – 0.6310 in)
Main bearing oil clearance .... 0.020 – 0.044 mm (0.00079 – 0.00173 in)

Clutch

Type .... Wet, multiplate
Number of plain plates:
- All models except XS250 SE (4G6), C (5H8) 6
- XS250 SE, (4G6), C (5H8) 4
Number of friction plates:
- All models except XS250 SE (4G6), C (5H8) 7
- XS250 SE, (4G6), C (5H8) 5
Plain plate thickness .... 1.6 ± 0.1 mm (0.063 ± 0.004 in)
Warpage limit .... 0.05 mm (0.002 in)
Friction plate thickness .... 3.0 mm (0.12 in)
Service limit .... 2.7 mm (0.11 in)
Number of springs 4
Spring free length:
- All models except XS250 SE (4G6), C (5H8) .... 34.6 mm (1.362 in)
- Service limit .... 33.6 mm (1.323 in)
- XS250 SE (4G6), C (5H8) .... 44.0 mm (1.732 in)
- Service limit .... 43.0 mm (1.693 in)

Gearbox

Type .... 6-speed constant mesh
Ratios:
- 1st 2.500:1 (35/14)
- 2nd 1.777:1 (32/18)
- 3rd 1.380:1 (29/21)
- 4th 1.125:1 (27/24)
- 5th 0.961:1 (25/26)
- Top 0.866:1 (26/30)
Secondary reduction ratio (UK models):
- XS400 (3N7) 2.438:1 (39/16)
- XS250 (3N6), (4A2) 2.933:1 (44/15)
- XS400 SE (4G5) 2.250:1 (36/16)
- XS250 SE (4G6) 2.733:1 (41/15)
- XS250 C (5H8) 2.438:1 (39/16)
Secondary reduction ratio (US models):
- XS400 E (2LO), F (2LO), 2F (2V6) 2.437:1 (39/16)
- All other models 2.312:1 (37/16)

Torque wrench settings

Component	kgf m	lbf ft
Cylinder head nuts:		
8 mm	2.2	16.0
6 mm	1.0	7.0
Primary drive sprocket	6.5	47.0
Crankcase bolts:		
8 mm	2.2	16.0
6 mm	1.0	7.0

Specifications relating to Chapter 2

Fuel tank
Overall capacity:
   XS400 (3N7), XS250 (3N6), XS250 (4A2), XS250 C
   (5H8) .... 17.0 litre (3.74 Imp gal)
   XS400 2E (2G5) .... 11.0 litre (2.91 US gal)
   Other models .... 14.0 litre (3.08/3.67 Imp/US gal)
Reserve capacity:
   XS400 G (3F8), SG (3F9), H (4R5), SH (4R4),
   SJ (14V) .... 3.30 litre (0.87 US gal)
   Other models .... Not available

Carburettors
Make .... Mikuni
Type:
   XS400 models BS34
   XS250 models BS32
ID mark:
   XS400 SE (4G5) 3X800
   XS250 SE (4G6) 4E000
   XS250 C (5H8) 1M001
   XS400 E (2LO), F (2LO), 2F (2V6) 2LO-60
   XS400 2E (2G5) 2G5-60
   XS400 G (3F8), SG (3F9), H (4R5), SH (4R4),
   SJ (14V) 3F9-00
   Other XS400 models 2G560
   Other XS250 models 1M000
Main jet:
   XS400 SE (4G5) 125.0
   XS250 SE (4G6) 110.0
   XS400 E (2LO), F (2LO), 2F (2V6) 132.5
   XS400 G (3F8), SG (3F9), H (4R5), SH (4R4),
   SJ (14V) 135.0
   Other XS400 models 137.5
   Other XS250 models 117.5
Air jet:
   XS250 models 0.6
   XS400 models 45
Jet needle:
   XS400 SE (4G5) 5HZ17
   XS250 SE (4G6) 4HX15
   XS400 G (3F8), SG (3F9), H (4R5), SH (4R4),
   SJ (14V) 5GZ9
   Other XS400 models 5Z1
   Other XS250 models 4Z1
Needle clip position:
   XS400 SE (4G5) .... 4th groove
   XS250 SE (4G6) .... 2nd groove
   XS400 G, (3F8), SG (3F9), H (4R5), SH (4R4),
   SJ (14V) .... Not Available
   Other XS400 models .... 3rd groove
   Other XS250 models .... 4th groove
Needle jet:
   XS400 SE (4G5) Y-2
   XS250 SE (4G6) Y-6
   XS400 G (3F8), SG (3F9), H (4R5), SH (4R4),
   SJ (14V) Y-2
   Other XS400 models X-6
   Other XS250 models X-8
Throttle valve:
   XS400 G (3F8), SG (3F9), H (4R5), SH (4R4),
   SJ (14V) .... Not Available
   Other XS400 models 135
   Other XS250 models 125
Pilot jet:
   XS250 SE (4G6) 17.5
   XS400 models 42.5
   Other XS250 models 20.0
Pilot air screw (turns out):
   XS400 SE (4G5) .... $3\frac{1}{2} \pm \frac{1}{2}$
   XS250 SE (4G6) .... $2\frac{3}{4} \pm \frac{1}{2}$

XS400 2E (2G5), G (3F8), SG (3F9), H (4R5), SH (4R4),
SJ (14V) .... Not available (preset)
   Other XS400 models .... $1\frac{1}{4} \pm \frac{1}{2}$
   Other XS250 models .... $1\frac{1}{2} \pm \frac{1}{2}$
Starter jet:
   XS400 SE (4G5), G (3F8), SG (3F9), H (4R5), SH (4R4),
   SJ (14V) 35
   Other XS400 models 30
   All XS250 models 25
Fuel level:
   XS250 C (5H8), XS400 E (2L0), F (2L0), 2F (2V6) .... 32 ± 1 mm (1.26 ± 0.04 in)
   XS400 H (4R5), SH (4R4), SJ (14V) .... 3.0 ± 1.0 mm (0.12 ± 0.004 in)
   Other models .... Not Available
Float height:
   XS400 G (3F8), SG (3F9) .... 27.3 ± 0.5 mm (1.07 ± 0.019 in)
   Other XS400 models and all XS250 models .... 25.7 ± 1.0 mm (1.01 ± 0.04 in)
Vacuum difference between cylinders .... 5 mm Hg or less
Idle speed .... 1200 ± 50 rpm

Engine lubrication
Oil capacity:
   Dry .... 2.6 litre (4.6/5.5 Imp/US pint)
   Oil change .... 2.0 litre (3.5/4.2 Imp/US pint)
   Oil and filter change .... 2.3 litre (4.2/4.9 Imp/US pint)
Oil grade:
   XS400 2E (2G5) .... Shell X-100 or Yamalube 4-cycle
   Other models:
      Above 5°C (41°F) .... SAE 20W/40 type SE motor oil
      Below 15°C (59°F) .... SAE 10W/30 type SE motor oil

Oil pump
Type .... Trochoid
End clearance .... 0.10 – 0.18 mm (0.004 – 0.007 in)
Tip clearance .... 0.03 – 0.09 mm (0.001 – 0.004 in)
Side clearance .... 0.03 – 0.09 mm (0.001 – 0.004 in)
Pump capacity @ 500 rpm .... 1.2 litre (2.1/2.5 Imp/US pint)
Relief valve opens at .... 5.0 ± 0.5 kg/cm^2 (71 ± 7 psi)
Bypass valve opens at .... 1.0 ± 0.2 kg/cm^2 (14 ± 3 psi)

Specifications relating to Chapter 3

Ignition system
Type:
   All UK models .... Coil and contact breaker
   XS400 G (3F8), SG (3F9), H (4R5), SH (4R4),
   SJ (14V) .... Capacitor discharge ignition (CDI)
   All other US models .... Coil and contact breaker

Ignition timing
Retarded:
   All UK models .... 10° BTDC
   All US models .... 10° BTDC @ 1200 rpm
Advanced:
   All UK models .... Not Available
   XS400 G (3F8), SG (3F9), H (4R5), SH (4R4),
   SJ (14V) .... 36° – 40° @ 2700 – 3400 rpm
   All other US models .... Not Available

Contact breaker
Contact gap .... 0.30 – 0.40 mm (0.012 – 0.016 in)
Dwell angle .... 105°

Ignition coil
Minimum spark gap .... 6.0 mm (0.24 in) or more @ 500 rpm
Primary winding resistance:
   XS400 G (3F8), SG (3F9), H (4R5), SH (4R4),
   SJ (14V) .... 3.0 ohms $\pm$ 10% @ 20°C (68°F)
   All other models .... 4.0 ohms $\pm$ 10% @ 20°C (68°F)
Secondary winding resistance:
   XS400 G (3F8), SG (3F9), H (4R5), SH (4R4),
   SJ (14V) .... 8.6 K ohms $\pm$ 20% @ 20°C (68°F)
   All other models .... 9.5 K ohms $\pm$ 20% @ 20°C (68°F)

Recommended spark plugs

Make:	
UK models	NGK BP7ES
US models	NGK BP7ES or Champion N7Y
Electrode gap	0.7 -- 0.8 mm (0.028 -- 0.032 in)

Alternative spark plugs (UK models)

Make	Champion
Type:	
All XS400 models	N7Y, N7YC or N7GY
All XS250 models	N7Y, N7YC or N7GY
Gap	0.7 mm (0.028 in)

Specifications relating to Chapter 4

Front forks

Type	Hydraulically-damped telescopic
Travel:	
XS250 (4A2)	Not Available
All other models	140 mm (5.5 in)
Fork spring free length:	
XS250 (4A2)	Not Available
All other UK models	502 mm (19.8 in)
XS400 G (3F8), 2G (3F9), H (4R5), SH (4R4) and SJ (14V)	502 mm (19.8 in)
All other US models	484 mm (19.02 in)
Fork stanchion OD:	
XS250 (4A2)	Not Available
All other UK models	33 mm (1.3 in)
XS400 G (3F8), 2G (3F9), H (4R5), SH (4R4) and SJ (14V)	Not Available
All other US models	33 mm (1.3 in)
Damping oil grade:	
All UK models	SAE 10W/30 type SE motor oil
XS400 2E (2G5)	Yamaha fork oil (20Wt) or SAE 20W motor oil
XS400 G (3F8), 2G (3F9), H (4R5), SH (4R4) and SJ (14V)	Yamaha fork oil (20Wt) or equivalent
All other US models	SAE 10W/30 type SE motor oil or Yamaha (or equivalent) fork oil
Quantity per leg:	
XS250 (4A2)	130 cc (4.6 Imp fl oz)
All other UK models	142 cc (4.9 Imp fl oz)
XS400 D (2A2) and 2E (2G5)	130 cc (4.4 US fl oz)
All other US models	142 cc (4.8 US fl oz)

Rear suspension

Type	Swinging arm, controlled by two hydraulically-damped coil spring suspension units
Suspension unit travel:	
XS250 (4A2)	Not Available
All other models	80 mm (3.15 in)
Spring free length:	
XS250 (4A2)	Not Available
All other UK models	216 mm (8.50 in)
XS400 D (2A2) and 2E (2G5)	205 mm (8.07 in)
All other US models	216 mm (8.50 in)
Swinging arm free play (service limit):	
XS250 (4A2)	Not Available
All other models	1.00 mm (0.04 in)
Swinging arm pivot shaft OD:	
XS250 (4A2)	Not Available
All other models	16.00 mm (0.63 in)

Torque wrench settings

Component	kgf m	lbf ft
Bottom yoke pinch bolt	3.5	25.5
Upper yoke pinch bolt	1.2	8.5
Rear suspension unit:		
Upper mounting	3.0	21.5
Lower mounting	3.0	21.5

Component	kgf m	lbf ft
Pivot shaft nut	6.5	47.0
Front wheel spindle	10.5	76.0
Front wheel spindle clamp	2.0	14.5
Rear wheel spindle	10.5	76.0

Specifications relating to Chapter 5

Wheels

	Front	Rear
Type:		
XS250 C (5H8), XS400 2E (2G5), 2F (2V6), G (3F8), H (4R5)	Wire spoked	Wire spoked
All other models	Cast aluminium	Cast aluminium
Size:		
XS400 (3N7), XS250 (3N6), XS250 (4A2)	1.85 x 18	2.15 x 18
XS400 SE (4G5), XS250 SE (4G6)	1.85 x 18	2.50 x 16
XS250 C (5H8)	1.60 x 18	1.85 x 18
XS400 D (2A2), E (2L0), F (2L0)	1.85 x 18	2.15 x 18
XS400 2E (2G5)	1.60 x 18	1.80 x 18
XS400 2F (2V6)	1.60 x 18	1.85 x 18
All other US models	Not Available	Not Available
Rim runout (service limit):		
Radial and axial	2.00 mm (0.08 in)	

Front brake

Type:	
XS400 2E (2G5), 2F (2V6), G (3F8), H (4R5)	Twin leading shoe (tls) drum
XS250 C (5H8)	Twin leading shoe (tls) drum
All other models	Single hydraulic disc brake
Disc brake:	
Disc diameter:	
All UK models	267 mm (10.5 in)
XS400 SG (3F9), SH (4R4), SJ (14V)	267 mm (10.5 in)
All other US models	257 mm (10.1 in)
Disc thickness	5.0 mm (0.2 in)
Service limit	4.5 mm (0.18 in)
Pad thickness:	
XS400 (3N7), XS250 (3N6), XS250 (4A2)	11.0 mm (0.24 in)
XS400 SE (4G5), XS250 SE (4G6)	10.2 mm (0.40 in)
Service limit:	
XS400 (3N7), XS250 (3N6), XS250 (4A2)	5.00 mm (0.20 in)
XS400 SE (4G5), XS250 SE (4G6)	5.70 mm (0.22 in)
Pad friction material thickness -- US models	6.5 mm (0.26 in)
Service limit – US models	1.5 mm (0.06 in)
Master cylinder ID	14.0 mm (0.55 in)
Caliper ID:	
XS400 (3N7), XS250 (3N6), XS250 (4A2)	38.1 mm (1.50 in)
XS400 SE (4G5), XS250 SE (4G6)	42.85 mm (1.69 in)
XS400 SG (3F9), SH (4R4), SJ (14V)	42.85 mm (1.69 in)
XS400 D (2A2), E (2L0), F (2L0)	38.1 mm (1.50 in)
Hydraulic fluid type	DOT 3 (US) SAE J1703 (UK)
Drum brake:	
Drum diameter	180 mm (7.1 in)
Shoe width	30.0 mm (1.18 in)
Lining thickness	4.0 mm (0.16 in)
Service limit	2.0 mm (0.08 in)
Return spring free length	68 mm (2.68 in)

Rear brake

Type:	
All UK models	Single leading shoe (sls) drum
XS400 D (2A2), E (2L0), F (2L0)	Single hydraulic disc brake
All other US models	Single leading shoe (sls) drum
Drum brake:	
Drum diameter	160 mm (6.30 in)
Shoe width	30.0 mm (1.18 in)
Lining thickness	4.0 mm (0.16 in)
Service limit	2.0 mm (0.08 in)
Disc brake:	
Master cylinder ID	15.8 mm (0.62 in)
Caliper ID:	
XS400 D (2A2), E (2L0), F (2L0)	38.1 mm (1.50 in)
Hydraulic fluid type	DOT 3 (US)

Front tyre
Size 3.00S18-4PR
Type:
    XS400 SG (3F9), SH (4R4), SJ (14V) .... Tubeless
    All other models .... Tubed
Pressures (cold) -- UK models:
    Up to 90 kg (198 lb) load .... 26 psi (1.8 kg/cm^2)
    90 -- 115 kg (198 -- 254 lb) load .... 28 psi (2.0 kg/cm^2)
    Continuous high-speed use .... 28 psi (2.0 kg/cm^2)
Pressures (cold) -- US models:
    XS400 F (2L0), 2F (2V6), G (3F8), SG (3F9), H (4R5),
    SH (4R4), SJ (14V):
        Up to 90 kg (198 lb) load .... 26 psi (1.8 kg/cm^2)
        90 -- 115 kg (198-254 lb) load .... 28 psi (2.0 kg/cm^2)
        Continuous high-speed use .... 28 psi (2.0 kg/cm^2)
    All other models:
        Normal riding .... 26 psi (1.8 kg/cm^2)
        High speed riding/passenger .... 28 psi (2.0 kg/cm^2)

Rear tyre
Size:
    XS400 (3N7), XS250 (3N6), XS250 (4A2), XS250 C
    (5H8) 3.75S18-4PR
    XS400 SE (4G5), XS250 SE (4G6) .... 120/90 -- 16 63S
    XS400 G (3F8), SG (3F9), H (4R5), SH (4R4),
    SJ (14V) .... 120/90 -- 16 63S
    XS400 D (2A2), E (2L0), 2E (2G5), F (2L0), 2F (2V6) ... 3.50S18-4PR
Type:
    XS400 SG (3F9), SH (4R4), SJ (14V) .... Tubeless
    All other models .... Tubed
Pressures (cold) -- UK models
    Up to 90 kg (198 lb) load .... 28 psi (2.0 kg/cm^2)
    90 -- 115 kg (198 -- 254 lb) load .... 32 psi (2.3 kg/cm^2)
    Continuous high-speed use .... 32 psi (2.3 kg/cm^2)
Pressures (cold) -- US models:
    XS400 F (2L0), 2F (2V6), G (3F8), SG (3F9), H (4R5),
    SH (4R4), SJ (14V):
        Up to 90 kg (198 lb) load .... 28 psi (2.0 kg/cm^2)
        90 -- 115 kg (198 -- 254 lb) load .... 32 psi (2.3 kg/cm^2)
        Continuous high-speed use .... 32 psi (2.3 kg/cm^2)
    All other models:
        Normal riding .... 28 psi (2.0 kg/cm^2)
        High speed riding/passenger .... 33 psi (2.3 kg/cm^2)

Torque wrench settings

Component	kgf m	lbf ft
Front disc brake:		
Disc mounting bolts	2.0	14.5
Caliper to bracket	1.8	13.0
Caliper fixed pad	0.3	2.0
Caliper bleed screw	0.6	4.5
Caliper bracket to fork	3.5	25.5
Brake hose union	2.6	19.0
Front wheel spindle	10.5	76.0
Front wheel spindle clamp	2.0	14.5
Rear wheel spindle	10.5	76.0

Specifications relating to Chapter 6

Battery
Make .... Yuasa. FB or GS
Model:
    XS250 C (5H8) 12N7-3B
    All other models 12N12-4A-1
Capacity:
    XS250 C (5H8) .... 12 volt 7 Ah
    All other models .... 12 volt 12 Ah
Charging rate:
    XS250 C (5H8) .... 0.7 A for 10 hours
    All other models .... 1.2 A for 10 hours
Earth (ground) .... Negative (--)

Alternator
Voltage .... 12 volt
Charging output .... 14.5V, 13A @ 5000 rpm

Field coil resistance	4.04 ohms ± 10% @ 20°C (68°F)
Stator coil resistance	0.72 ohms ± 10% @ 20°C (68°F)

Rectifier

Type	6 element, full wave
Make	Mitsubishi or Stanley
Model	DS10TEY or DE3804-1
Withstand voltage	400V
Operating voltage	12V

Voltage regulator

All UK models and US XS400 F (2L0), 2F (2V6), G (3F8), SG (3F9), H (4R5), SH (4R4), SJ (14V):

Type	Electronic
Make	Nippon Denso
Model	026000-3280
Regulated voltage:	
XS400 H, (4R5), SH (4R4), SJ (14V)	14.35 ± 0.35V
All other models	14.5 ± 0.5V

XS400 D (2A2), E (2L0), 2E (2G5):

Type	Electro-mechanical
Core gap	0.2 mm (0.008 in) minimum
Point gap	0.1 mm (0.004 in) minimum
Voltage coil resistance	10.4 – 10.7 ohms @ 20°C (68°F)
Resistor resistance	140 ohms ± 10% @ 20°C (68°F)

Starter motor

Type	Constant mesh
Make	Mitsuba
Model	SM223B
Output	0.5 kW
Armature winding resistance	0.005 ohms ± 10% @ 20°C (68°F)
Field coil resistance	0.011 ohms ± 10% @ 20°C (68°F)
Brush length	11.0 – 12.5 mm (0.43 – 0.49 in)
Service limit	6.0 mm (2.4 in)
Commutator OD	28 mm (1.10 in)
Service limit	27 mm (1.06 in)
Mica undercut	0.7 mm (0.03 in)

Starter relay

Type	Electromagnetic
Make	Hitachi
Model	A104-70
Contact rating	100A
Cut-in voltage	6.5V or less
Winding resistance	3.5 ohms ± 10%

Horn

Maximum amperage	2.5A

Turn signal relay

Frequency	85 ± 10 cycles per minute
Capacity	27W x 2 + 3.4W

Fuse ratings

Main (red)	20A
Headlamp (red/yellow)	10A
Signal (brown)	10A
Ignition (red/white)	10A

Bulb wattages

Headlamp:

XS250 (3N6), XS400 (3N7), XS250 C (5H8)	12V 50/40W
XS250 SE (4G6), XS400 SE (4G5)	12V 45/45W
XS400 D (2A2), E (2L0), 2E (2G5), F (2L0), 2F (2V6)	12V 40/30W
XS400 G (3F8), SG (3F9), H (4R5), SH (4R4), SJ (14V)	12V 50/35W

Turn signal lamps:

XS250 C (5H8)	12V 21W
All other models	12V 27W

Tail/brake lamp:
 UK models 12V 5/21W
 US models 12V 8/27W
Warning lamps:
 Turn signal 12V 3.4W
 Neutral 12V 3.4W
 High beam 12V 3.4W
 Oil pressure 12V 3.4W
Instrument lamps 12V 3.4W
Parking lamp -- UK only 12V 3.4W
Number (license) plate lamp:
 XS250 SE (4G6), XS400 SE (4G5) 12V 3.4W
 XS400 G (3F8), SG (3F9), H (4R5), SH (4R4),
 SJ (14V) 12V 3.8W

1 Introduction

This Chapter covers the models produced since the original manual was published, providing specifications and working procedures where different from that given in Chapters 1 to 6. Most changes relate to the fitting of different cycle parts. A number of relatively minor mechanical modifications were incorporated, and certain assemblies, such as the brake calipers and wheels, were altered as newer designs became available.

To facilitate model identification a summary is given at the end of this Section. In each case, the model has a code number, eg 3N6. This allows a specific machine to be distinguished from an earlier or later model and is thus useful when ordering parts for the machine.

Where possible, the month and year of introduction and discontinuation are shown, although where these are not available the year only is shown. Note that there is usually an overlap between the introduction of a new model and the phasing out of its predecessor, and that old models can remain unsold in dealers' showrooms for several months after the official date of their demise.

UK model summary

Model	Code	Introduced	Discontinued
XS250 P*	1U5	January 1977	June 1979
XS250	3N6	January 1978	During 1980
XS250	4A2	During 1980	During 1981
XS250 SE	4G6	May 1980	During 1984
XS250 C	5H8	February 1981	July 1982
XS400*	2JO	September 1977	During 1979
XS400	3N7	During 1979	April 1980
XS400 SE	4G5	March 1980	October 1983

Refer to Chapters 1 to 6

US model summary

Model	Code	Introduced	Model year
XS360 C*	1L9	September 1975	1976
XS360 2D*	1T6	October 1976	1977
XS400 D	2A2	October 1976	1977
XS400 E	2L0	September 1977	1978
XS400 2E	2G5	September 1977	1978
XS400 F	2L0	September 1978	1979
XS400 2F	2V6	September 1978	1979
XS400 G Special II	3F8	December 1979	1980
XS400 SG Special	3F9	November 1979	1980
XS400 H Special II	4R5	September 1980	1981
XS400 SH Heritage	4R4	September 1980	1981
XS400 SJ Heritage	14V	October 1981	1982

* Refer to Chapters 1 to 6

2 Model development

UK models

XS250 (3N6) and XS400 (3N7) models

These models received a number of detail changes, mostly of a cosmetic nature. These included a new paint scheme to suit the new, larger, fuel tank, and a passenger grab rail was added. On a more practical level, the seat height was reduced by $1\frac{1}{4}$ in (35 mm) and the front footrests moved back by 3.9 in (100 mm). The latter alteration necessitated the introduction of a remote gearchange linkage and alterations to the brake pedal and main stand positions.

A revised exhaust system was fitted to comply with the repositioned foot controls and to complement the revised styling. The swinging arm assembly was lengthened, and this, combined with a change in both the front and rear suspension spring rates, was designed to improve stability and handling. At the rear of the machine, a drum brake was fitted in place of the previous disc arrangement. The steering lock was integrated with the ignition lock, and the instruments were modified, the new units featuring orange-tinted internal illumination. Other electrical modifications included a revised ignition coil.

XS250 (4A2) model

The XS250 (4A2) model was introduced during 1980 as a successor to the previous (3N6) model. It differed only in detail from its predecessor.

XS250 SE (4G6) and XS400 SE (4G5) US Custom/Special models

These models reflect the popular 'factory custom' trend. The main changes were cosmetic, notably the fitting of a stepped dualseat and a teardrop-shaped fuel tank, a revised exhaust system with a balance pipe and short silencers, and high, pullback handlebars. Incidental to the styling alterations was the repositioning of the footrests and the associated controls; these were moved forward and the remote gearchange linkage dropped.

Mechanical changes included a revised clutch assembly, the fitting of an oil level sight glass in place of a dipstick (1981 onwards) and modified engine mountings. The most significant change to the cycle parts was the fitting of a new front brake caliper, and a sixteen inch rear wheel with a 'fat', high-profile tyre.

XS250 C (5H8) model

The XS250C is an economy model, featuring many of the mechanical modifications of the XS250 SE, such as the revised clutch and the oil level sight glass. As an economy machine, the starter motor was omitted, starting being by kickstart only. Wire spoked wheels were fitted in place of the cast alloy items, the front hub incorporating a twin leading shoe drum brake.

US models

XS400 E (2L0) and XS400 2E (2G5)

Basically similar to the XS400 D (2A2) model which preceded it, the XS400 E model featured detail styling and mechanical changes. Minor engine revisions included restyling of the cylinder head and barrel castings. A revised seat and tail lamp was fitted and the fuel tap was modified to include a drain plug. Carburettor changes included the adoption of a two-stage cold-start circuit. The swinging arm pivot bolt was modified to include a grease nipple at each end. Revised instruments and headlamp mountings were incorporated. The electro-mechanical voltage regulator was replaced by an electronic version.

The XS400 2E model is an economy model and is not fitted with a centre stand, starter motor or turn signal self-cancelling unit. The cast wheels and disc brakes of the E model were replaced by wire spoked wheels and drum brakes.

XS400 F (2L0) and XS400 2F (2V6)

The 1979 F models were very similar to the 1978 E machines, featuring minor revisions and refinements.

XS400 G Special II (3F8) and XS400 SG Special (3F9)

These were the 1980 models and were restyled to conform with the factory custom image which had become very popular. The SG was the full equipment version, while the G, or Special II, was an economy model. Engine modifications were minor, such as the fitting of an oil level sight glass plus detail specification changes. The coil-and-contact breaker ignition system used on all previous machines was replaced by a capacitor discharge electronic system, the pickup assembly being concealed behind a cover on the cylinder head. This last (emission law related) feature was accompanied by the modified carburettors with non-adjustable pilot screws.

The revised styling included a stepped seat, pullback handlebars and a new exhaust system. This featured a balance pipe and short, megaphone-type silencers. The rear wheel size was reduced to 16 inch to allow the fitting of a 'fat', high-profile rear tyre. Wheels were wire-spoked with drum brakes for the G model, and cast alloy with front disc and rear drum in the case of the SG.

XS400 H Special II (4R5) and XS400 SH Heritage (4R4)

For 1981, the Special II economy-custom was continued, the previous years Special being renamed as the Heritage for this year. No significant changes were introduced, apart from revised engine mountings.

XS400 SJ Heritage (14V)

The last of the sohc range, the 1982 Heritage inherited the basic equipment of the 1981 Special II, namely the wire spoked wheels and drum brakes. After 1982 the sohc twins were discontinued in favour of the dohc-engined machines.

3 Gearchange linkage: adjustment – XS250 (3N6), (4A2) and XS400 (3N7) UK models

1 On the above models, the footrest position was altered to give an improved riding position. Moving the footrests back necessitated altering the location of the gearchange pedal, and this was accomplished by fitting a remote linkage arrangement. The new pedal projects rearwards from its pivot, movement being transmitted through an adjustable rod and pivots to a short gearchange arm clamped to the splined gearchange shaft.

2 To remove the linkage, without disturbing the adjustment, prise off the circlip which retains the gearchange pedal to its pivot and remove the gearchange arm pinch bolt. The assembly can then be slid off.

3 When refitting the assembly, select neutral and ensure that the slot in the gearchange arm is aligned with the '4' cast in the crankcase cover. The slot should be positioned horizontally, with the arm running forward of the shaft. This initial setting should be checked as part of the adjustment procedure,

especially if it is suspected that the setting is incorrect. It is also advisable to dismantle the linkage before making adjustment so that the pedal pivot can be lubricated with engine oil. Once in position, tighten the pinch bolt and refit the circlip which retains the pedal to its pivot.

4 The pedal position can be adjusted by turning the rod between the gearchange arm and pedal. Start by slackening the locknuts at each end of the rod, noting that one end is marked 'L' to denote a left-hand thread. Turn the rod either way as necessary, until the top of the pedal rubber lies approximately 16 mm (0.63 in) below the top of the footrest rubber. This is an approximate setting only, and may be varied according to personal preference. Note, however, that if the angle between the rod and either the pedal or gearchange arm becomes acute, gearchanging will prove difficult, so adjustment should be kept to a minimum from the nominal setting. Once adjustment is correct, secure the locknuts.

4 Clutch: modifications – XS250 SE (4G6), C (5H8) UK models

The clutch differs in that it has fewer plain and friction plates. In consequence, the clutch spring lengths are altered, as will be seen from the specifications at the beginning of this Chapter. In other respects the clutch assembly is similar to the remaining models, the approach to dismantling and reassembly being unchanged.

5 Oil level sight glass: general – XS250 SE (4G6), XS250 C (5H8), XS400 SE (4G5) UK models, and XS400 G, SG, H, SH, SJ US models

The above models were fitted with an oil level sight glass in the right-hand crankcase cover, in place of the dipstick used on earlier models. This simplifies regular checking of the engine oil level. To check the oil level, run the engine for several minutes, then switch the engine off and allow the oil level to settle. With the machine standing vertically on level ground, check that the oil level lies between the minimum and maximum marks; top up with the recommended oil if required.

6 Fuel tap: modification

1 Most later models are equipped with a revised fuel tap incorporating a drain bolt. This allows the tap to be drained and any accumulated sediment or water droplets to be flushed out of the fuel system. When working on the fuel system never smoke and take all suitable precautions to prevent the risk of fire. When removing the drain bolt, set the tap to the 'ON' or 'RES' positions to avoid large quantities of fuel being spilt. Unscrew the bolt and check for signs of water or sediment contamination.

2 If required, the tank can be drained or the tap flushed by turning the tap lever to the 'PRI' position and draining the fuel into a clean metal container. If significant amounts of water or dirt are expelled it will probably be preferable to remove the tank for more thorough cleaning. Remember that residual dirt or water may remain in the carburettors, and it may prove necessary to remove and dismantle these too.

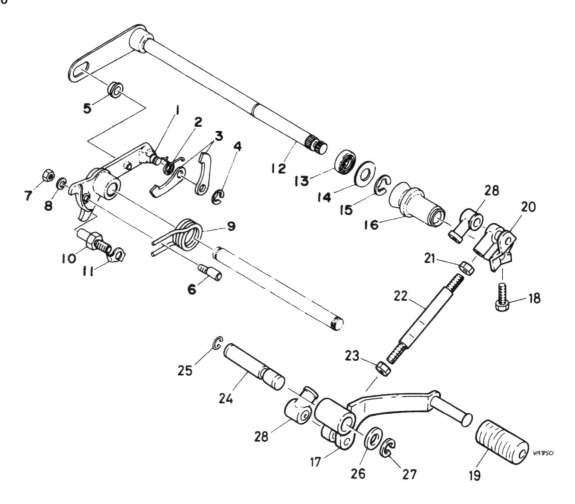

Fig. 7.1 Gearchange linkage

1 Gear selector arm	8 Spring washer	15 Circlip	22 Linkage
2 Pawl spring	9 Return spring	16 Sealing rubber	23 Nut
3 Gearchange pawl	10 Stop	17 Gearchange lever	24 Lever pivot
4 Circlip	11 Lock washer	18 Pinch bolt	25 Circlip
5 Roller	12 Gearchange shaft	19 Lever rubber	26 Washer
6 Centraliser screw	13 Oil seal	20 Short gearchange arm	27 Circlip
7 Nut	14 Washer	21 Nut	28 Boot – 2 off

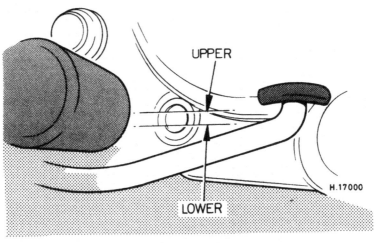

Fig. 7.2 Engine oil level sight glass

7 Air filter: modification

The air filter arrangement was modified on the later machines, making access for cleaning and renewal much easier. The separate casings were combined into a single unit, each element being housed behind a flat cover. To gain access to the elements, pull off the side panels, then remove the two screws holding each cover to the casing. Lift away the covers to reveal the elements. Refer to Chapter 2, Section 9 for details of element cleaning. Refitting is a straightforward reversal of the removal sequence.

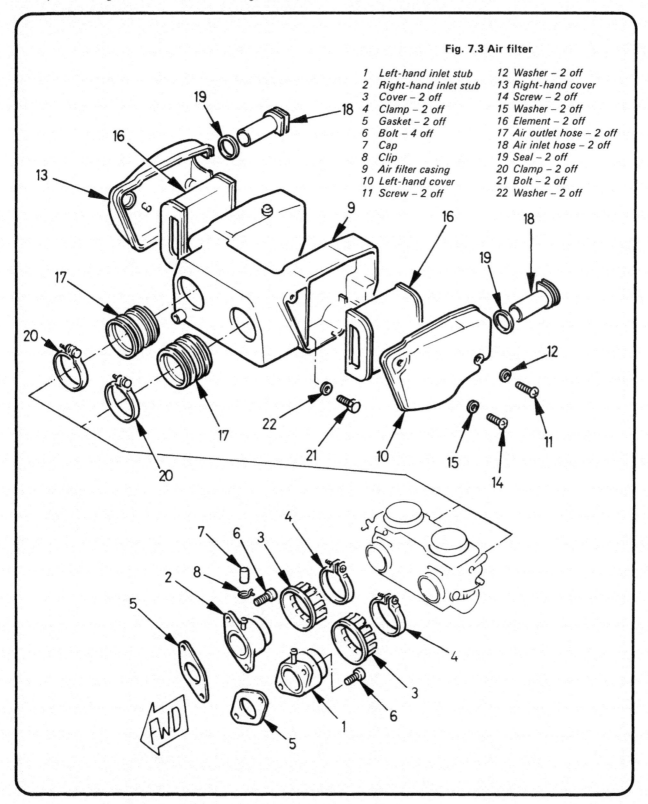

Fig. 7.3 Air filter

1	Left-hand inlet stub	12	Washer – 2 off
2	Right-hand inlet stub	13	Right-hand cover
3	Cover – 2 off	14	Screw – 2 off
4	Clamp – 2 off	15	Washer – 2 off
5	Gasket – 2 off	16	Element – 2 off
6	Bolt – 4 off	17	Air outlet hose – 2 off
7	Cap	18	Air inlet hose – 2 off
8	Clip	19	Seal – 2 off
9	Air filter casing	20	Clamp – 2 off
10	Left-hand cover	21	Bolt – 2 off
11	Screw – 2 off	22	Washer – 2 off

8 Fuel level: measurement – XS400 H, SH and SJ US models

1   These models are equipped with a drain screw on each float bowl which doubles as an adaptor for checking the fuel level. This allows the fuel level to be checked without having to dismantle the carburettors, as would be necessary to check float height adjustment. Note, however, that if adjustment proves necessary, dismantling will still be required. To carry out the check, a length of clear tubing with an inside diameter of 6 mm (0.24 in) will be required.

2   Place the machine on its centre stand on level ground, then use a jack or similar to position the machine so that the carburettors are vertical. Connect the tubing to one of the drain screws and position the open end so that it is well above the level of the float bowl and arranged vertically alongside the float bowl gasket face. Turn the fuel tap to the 'ON' or 'RES' position, then open the drain screw to fill the tube with fuel. Start and run the engine for a few minutes to allow the fuel level to stabilise.

3   Stop the engine and note the position of the fuel in the tube in relation to the gasket face, then move the open end of the tube across to the other side of the machine, placing it alongside the second float bowl. Run the engine again, then switch off and note the fuel position. If there is a difference between the two fuel positions, adjust the angle of the machine by packing slips of wood or card beneath the centre stand feet as required. When the two levels are equal, the machine is positioned vertically and the fuel level test proper can be carried out.

4   Connect the tube to each drain screw in turn, opening the screw and allowing the fuel level to stabilise by running the engine, as described above. Switch off the engine, and measure the level of the fuel in the tube below the gasket face of the main carburettor body. Tighten the drain screw, connect the tube to the remaining float bowl and repeat the test on the second carburettor. In each case, the fuel level should be 3.0 $\pm$ 1.0 mm (0.12 $\pm$ 0.04 in) below the gasket face, and the level should be similar on both carburettors.

5   If the level is incorrect, or there is a significant difference between the two carburettors, remove them and detach the float bowls so that the floats, float valves and valve seats can be checked for wear or damage. If no problem is noted, adjust the fuel level by bending the tang on the float assembly by a small amount. Reassemble the carburettors, install them, and repeat the test as described above.

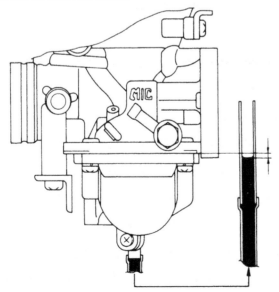

Fig. 7.4 Fuel level measurement

9 Electronic ignition system: general information

1   The US Special, Special II and Heritage models featured a capacitor discharge type ignition system, known by the manufacturer as TCI, in place of the coil and contact breaker arrangement used on previous models. The new arrangement provides more consistent and accurate ignition sparking, and by eliminating all mechanical contact, and thus wear, obviates the need for regular maintenance or renewal of ignition components.

2   The system is controlled by a pickup assembly comprising a baseplate carrying the pickup coils mounted on the left-hand side of the cylinder head, and a rotor attached to the end of the camshaft. As the camshaft rotates, the rotor sweeps past the pickup coils, generating a small trigger voltage as it does so. This signal is fed to the TCI unit, which contains the necessary ignition advance and amplifier circuits. The trigger signal causes the TCI unit to discharge its capacitors through the appropriate ignition coil, producing the high tension pulse which jumps across the spark plug electrodes.

3   Other circuitry in the TCI unit monitors the frequency of the trigger pulses as a means of gauging engine speed. This in turn allows the ignition to be advanced as engine speed rises. A subsystem within the TCI unit limits the duration of the primary ignition current flow to minimise electrical consumption of the system.

4   A secondary function of the TCI unit is to protect the ignition circuit; if the ignition switch is turned on, but the engine is not started, the absence of trigger pulses for more than a few seconds causes the TCI unit to switch off the supply to the ignition coils. This means that the ignition coil is effectively protected from damage due to overheating of the windings. As soon as the trigger pulses are sensed, the system is restored to normal operation.

10 Electronic ignition system: fault diagnosis

1   The TCI system should normally be trouble-free in operation, and as has been mentioned, requires no maintenance in the normal sense. In the event of an ignition fault developing, check through the system in the sequence outlined below, bearing in mind the following precautions: *Never run the engine without both spark plug caps connected; the high secondary voltage could damage the ignition coils. Beware of the risk of shocks from the ignition system. The voltage present in parts of the system is high enough to be very unpleasant or even dangerous. Carry out all checks with the engine at normal operating temperature, where possible. This ensures that faults which only become apparent at fairly high temperatures are highlighted.*

Test procedure	Action required
a) Ignition system wiring connections	Repair or renew any damaged wiring. Remake corroded or damaged connections. Check operation of the ignition switch and engine stop switch. Check for water contamination.
b) Battery condition	Make sure that electrolyte level is between maximum and minimum marks. Ensure that battery is fully charged.
c) Supply to system	Check fuse and fuse connections. Renew fuse and remake connections as required.
d) Ignition coils	Check primary and secondary winding resistances. If outside specifications, renew the coil(s).

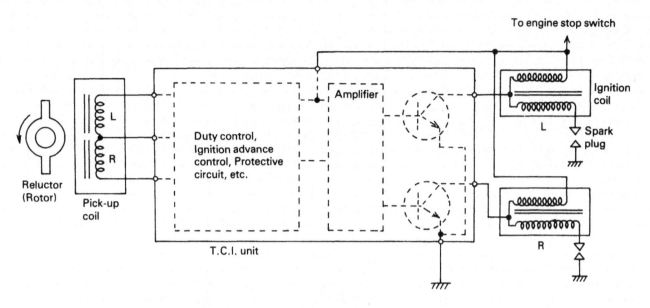

Fig. 7.5 Electronic ignition system circuit diagram

e) Pickup coils

Measure pickup coil resistance. If outside specified tolerance of 700 ± 150 ohms, renew the pickup coil assembly (see Section 11).

f) TCI unit

Test by substitution or have a full system test performed by a Yamaha dealer. If the unit is faulty it must be renewed; its sealed construction does not allow repairs to be made. Check local motorcycle breakers for a sound secondhand unit before buying a new one.

11 Electronic ignition system: pickup renewal

1   To prevent ignition timing alterations which might affect exhaust emissions, the pickup cover is held in place by screws, the heads of which are covered by blind anti-tamper plugs. To gain access to the pickup for renewal or in the course of an overhaul it is necessary to extract the plugs. This requires the use of a slide-hammer extractor, a centre punch, a drill, a 5 mm drill bit and a M6 x 1.0 mm plug tap. It should be noted that regular timing adjustment is not needed, so on the rare occasions that access to the pickup is required, many owners may prefer to have the plug removal operation carried out by a dealer. For those having the required equipment, proceed as follows.

2   Carefully centre punch the exact middle of each plug and drill a 5 mm pilot hole through each. Tap a M6 x 1.0 mm pitch thread in each plug, being extremely careful to avoid breaking the tap off in the plug; removal might prove very difficult if this were to occur. Screw the slide-hammer tool into each plug in turn and drive the plugs out of the cover.

3   With the anti-tamper plugs extracted, the cover screws can be removed and access gained to the pickup. If the pickup is disturbed or renewed, check the ignition timing using a stroboscopic timing lamp connected to the left-hand plug lead.

If possible, use a xenon tube type in preference to the cheaper neon versions. With the engine warmed up and running at 1200 rpm, the 'LF' mark on the rotor should align exactly with the fixed index mark. If this is not the case, slacken the baseplate screws and adjust the position of the pickup coils until the timing is correct. Tighten the screws and re-check the timing before refitting the cover.

4   Once the cover has been refitted, the screw heads must be sealed with new blind plugs. Note that in some areas, failure to do so may be deemed a violation of the emission laws. Tap the plugs squarely into the screw holes, noting that if they are fitted badly the cover is likely to crack, necessitating its renewal. Tap the plugs home until they are firmly located in the screw holes.

12 Frame: modifications

1   There were detail changes to the frames of later models, due mainly to the relocation of brackets and attachment points for the ancillary and electrical components. A more significant alteration was the modified engine mounting arrangement employed on the UK XS250 SE and XS400 SE, and on the US XS400 H, SH and SJ machines.

2   With the new arrangement, the previous rigid mounting system was replaced by an elastic type to isolate engine vibration more effectively. This entailed the modification of the engine attachment points on the frame and the addition of mounting plates for the front mounting point, and revised damper rubbers.

3   The change does not materially affect work on the machine; the method of engine removal being affected only in detail. The following torque wrench settings should be noted, however.

Component	kgf m	lbf ft
Upper mounting	1.8	13.0
Front mounting	3.0	21.7
Rear mounting	3.0	21.7

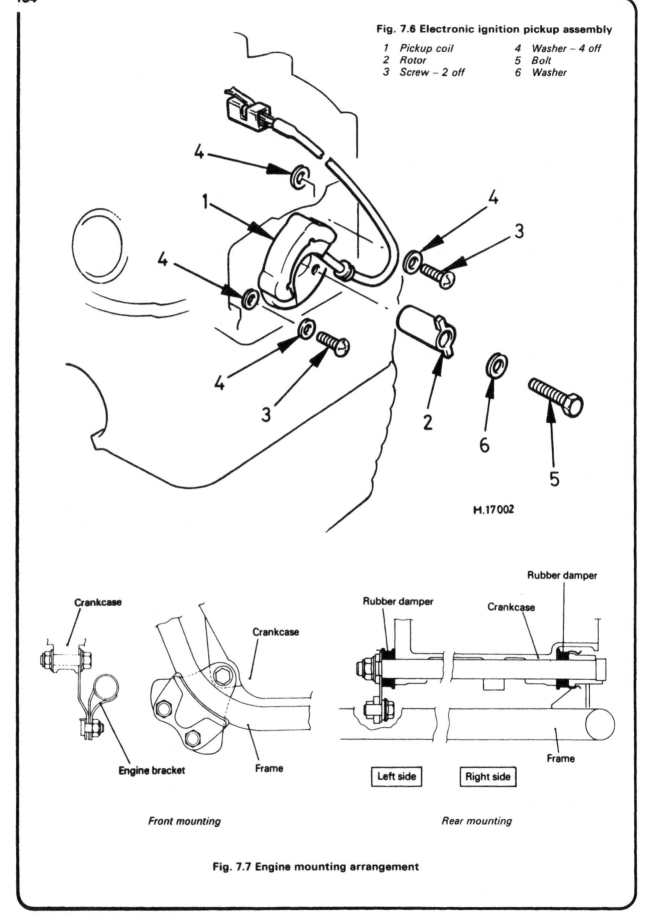

Fig. 7.6 Electronic ignition pickup assembly

1 Pickup coil
2 Rotor
3 Screw – 2 off
4 Washer – 4 off
5 Bolt
6 Washer

H.17002

Front mounting

Rear mounting

Fig. 7.7 Engine mounting arrangement

13 Front forks: modification

With the exception of detail changes, the front fork assembly remained unchanged throughout the model range. Variations occur due to different wheel fitments, the wheel spindle being fitted from the opposite side in some instances. It should be noted that the fork oil grade and quantity did vary, however, and information on this will be found in the Specifications that accompany this Chapter.

14 Swinging arm pivot: modification

The swinging arm pivot bolt was modified on later machines to provide better lubrication facilities for the pivot bushes. The revised pivot was fitted with a grease nipple at one end, and the nut was similarly equipped at the other. This allows grease to be pumped into the pivot at regular intervals without the need for removal. Greasing should be carried out annually, or every 5000 miles, whichever comes first, using a lithium-based grease.

15 Rear wheel: removal and refitting

1 The procedure for removing and refitting the rear wheel is largely the same as described in Chapter 5, Section 13, except that, with the exception of the US XS400 D, E and F models, the wheel is of the drum brake type. As with the earlier disc braked versions, it is necessary on most of the models covered in this Chapter to compress the rear suspension slightly, holding it in this position with the steel cable supplied in the machine's tool kit. Note also that the brake torque arm must be disconnected before the wheel can be removed.
2 The final drive chain fitted to some later models is of the endless type, no joining link being fitted. Inspect the chain closely, if no joining link is found the chain must be separated using a special chain breaking tool, and this should be available from most motorcycle dealers. If refitting an endless type chain, note that a new soft link will be needed each time. Many owners will have replaced the endless chain with a conventional spring link type, obviating the need for special tools or new links. When reconnecting this type of chain, ensure that the joining link clip is fitted so that its closed end faces in the direction of chain travel.

16 Tyres: removal and refitting – tubeless tyres

1 It is strongly recommended that should a repair to a tubeless tyre be necessary, the wheel is removed from the machine and taken to a tyre fitting specialist who is willing to do the job or taken to an official dealer. This is because the force required to break the seal between the wheel rim and tyre bead is considerable and considered to be beyond the capabilities of an individual working with normal tyre removing tools. Any abortive attempt to break the rim to bead seal may also cause damage to the wheel rim, resulting in an expensive wheel replacement. If, however, a suitable bead releasing tool is available, and experience has already been gained in its use, tyre removal and refitting can be accomplished as follows.
2 Remove the wheel from the machine by following the instructions for wheel removal as described in the relevant Section of this Chapter or Chapter 5. Deflate the tyre by removing the valve insert and when it is fully deflated, push the bead of the tyre away from the wheel rim on both sides so that the bead enters the centre well of the rim. As noted, this operation will almost certainly require the use of a bead releasing tool.
3 Insert a tyre lever close to the valve and lever the edge of

the tyre over the outside of the wheel rim. Very little force should be necessary; if resistance is encountered it is probably due to the fact that the tyre beads have not entered the well of the wheel rim all the way round the tyre. Should the initial problem persist, lubrication of the tyre bead and the inside edge and lip of the rim will facilitate removal. Use a recommended lubricant, a diluted solution of washing-up liquid or french chalk. Lubrication is usually recommended as an aid to tyre fitting but its use is equally desirable during removal. The risk of lever damage to wheel rims can be minimised by the use of proprietary plastic rim protectors placed over the rim flange at the point where the tyre levers are inserted. Suitable rim projectors may be fabricated very easily from short lengths (4-6 inches) of thick-walled nylon petrol pipe which have been split down one side using a sharp knife. The use of rim protectors should be adopted whenever levers are used and, therefore, when the risk of damage is likely.
4 Once the tyre has been edged over the wheel rim, it is easy to work around the wheel rim so that the tyre is completely free on one side.
5 Working from the other side of the wheel, ease the other edge of the tyre over the outside of the wheel rim, which is furthest away. Continue to work around the rim until the tyre is freed completely from the rim.
6 Refer to the following Section for details relating to puncture repair and the renewal of tyres. See also the remarks relating to the tyre valves in Section 18.
7 Refitting of the tyre is virtually a reversal of the removal procedure. If the tyre has a balance mark (usually a spot of coloured paint), as on the tyres fitted as original equipment, this must be positioned alongside the valve. Similarly, any arrow indicating direction of rotation must face the right way.
8 Starting at the point furthest from the valve, push the tyre bead over the edge of the wheel rim until it is located in the central well. Continue to work around the tyre in this fashion until the whole of one side of the tyre is on the rim. It may be necessary to use a tyre lever during the final stages. Here again, the use of a lubricant will aid fitting. It is recommended strongly that when refitting the tyre only a recommended lubricant is used because such lubricants also have sealing properties. Do not be over generous in the application of lubricant or tyre creep may occur.
9 Fitting the upper bead is similar to fitting the lower bead. Start by pushing the bead over the rim and into the well at a point diametrically opposite the tyre valve. Continue working round the tyre, each side of the starting point, ensuring that the bead opposite the working area is always in the well. Apply lubricant as necessary. Avoid using tyre levers unless absolutely essential, to help reduce damage to the soft wheel rim. The use of the levers should be required only when the final portion of bead is to be pushed over the rim.
10 Lubricate the tyre beads again prior to inflating the tyre, and check that the wheel rim is evenly positioned in relation to the tyre beads. Inflation of the tyre may well prove impossible without the use of a high pressure air hose. The tyre will retain air completely only when the beads are firmly against the rim edges at all points and it may be found when using a foot pump that air escapes at the same rate as it is pumped in. This problem may also be encountered when using an air hose on new tyres which have been compressed in storage and by virtue of their profile hold the beads away from the rim edges. To overcome this difficulty, a tourniquet may be placed around the circumference of the tyre, over the central area of the tread. The compression of the tread in this area will cause the beads to be pushed outwards in the desired direction. The type of tourniquet most widely used consists of a length of hose closed at both ends with a suitable clamp fitted to enable both ends to be connected. An ordinary tyre valve is fitted at one end of the tube so that after the hose has been secured around the tyre it may be inflated, giving a constricting effect. Another possible method of seating beads to obtain initial inflation is to press the tyre into the angle between a wall and the floor. With the airline attached to the valve additional pressure is then applied to the

tyre by the hand and shin, as shown in the accompanying illustration. The application of pressure at four points around the tyre's circumference whilst simultaneously applying the airhose will often effect an initial seal between the tyre beads and wheel rim, thus allowing inflation to occur.

11 Having successfully accomplished inflation, increase the pressure to 40 psi and check that the tyre is evenly disposed on the wheel rim. This may be judged by checking that the thin positioning line found on each tyre wall is equidistant from the rim around the total circumference of the tyre. If this is not the case, deflate the tyre, apply additional lubrication and reinflate. Minor adjustments to the tyre position may be made by bouncing the wheel on the ground.

12 Always run the tyre at the recommended pressures and never under or over-inflate. The correct pressures for various weights and configurations are given in the Specifications Section of this Chapter.

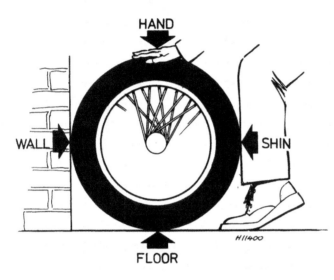

Fig. 7.8 Method of seating the beads on tubeless tyres

17 Puncture repair and tyre renewal – tubeless tyres

1 The primary advantage of the tubeless tyre is its ability to accept penetration by sharp objects such as nails etc without loss of air. Even if loss of air is experienced, because there is no inner tube to rupture, in normal conditions a sudden blow-out is avoided.

2 If a puncture of the tyre occurs, the tyre should be removed for inspection for damage before any attempt is made at remedial action. The temporary repair of a punctured tyre by inserting a plug from the outside should not be attempted. Although this type of temporary repair is used widely on cars, the manufacturers strongly recommend that no such repair is carried out on a motorcycle tyre. Not only does the tyre have a thinner carcass, which does not give sufficient support to the plug, the consequences of a sudden deflation is often sufficiently serious that the risk of such an occurrence should be avoided at all costs.

3 The tyre should be inspected both inside and out for damage to the carcass. Unfortunately the inner lining of the tyre – which takes the place of the inner tube – may easily obscure any damage and some experience is required in making a correct assessment of the tyre condition.

4 There are two main types of tyre repair which are considered safe for adoption in repairing tubeless motorcycle tyres. The first type of repair consists of inserting a mushroom-headed plug into the hole from the inside of the tyre. The hole is prepared for insertion of the plug by reaming and the application of an adhesive. The second repair is carried out by buffing the inner lining in the damaged area and applying a cold or vulcanised patch. Because both inspection and repair, if they are to be carried out safely, require experience in this type of work, it is recommended that the tyre be placed in the hands of a repairer with the necessary skills, rather than repaired in the home workshop.

5 In the event of an emergency, the only recommended 'get-you-home' repair is to fit a standard inner tube of the correct size. If this course of action is adopted, care should be taken to ensure that the cause of the puncture has been removed before the inner tube is fitted. It will be found that the valve in the rim is considerably larger than the diameter of the inner tube valve stem. To prevent the ingress of road dirt, and to help support the valve, a spacer should be fitted over the valve.

6 In the event of the unavailability of tubeless tyres, ordinary tubed tyres fitted with inner tubes of the correct size may be fitted. Refer to the manufacturer or a tyre fitting specialist to ensure that only a tyre and tube of equivalent type and suitability is fitted, and also to advise on the fitting of a valve nut to the rim hole.

18 Tyre valves: description and renewal – tubeless tyres

1 It will be appreciated from the preceding Sections, that the adoption of tubeless tyres has made it necessary to modify the valve arrangement, as there is no longer an inner tube which can carry the valve core. The problem has been overcome by fitting a separate tyre valve which passes through a close-fitting hole in the rim, and which is secured by a nut and locknut. The valve is fitted from the rim well, and it follows that the valve can be removed and replaced only when the tyre has been removed from the rim. Leakage of air from around the valve body is likely to occur only if the sealing seat fails or if the nut and locknut become loose.

2 The valve core is of the same type as that used with tubed tyres, and screws into the valve body. The core can be removed with a small slotted tool which is normally incorporated in plunger type pressure gauges. Some valve dust caps incorporate a projection for removing valve cores. Although tubeless tyre valves seldom give trouble, it is possible for a leak to develop if a small particle of grit lodges on the sealing face. Occasionally, an elusive slow puncture can be traced to a leaking valve core, and this should be checked before a genuine puncture is suspected.

3 The valve dust caps are a significant part of the tyre valve assembly. Not only do they prevent the ingress of road dirt into the valve, but also act as a secondary seal which will reduce the risk of sudden deflation if a valve core should fail.

19 Brake: modifications – general

1 It will be noted from the specifications that most of the later models departed from the earlier practice of fitting front and rear disc brakes to all cast wheel models; as a general rule all models were fitted with drum type rear brakes, whilst the twin leading shoe drum brake was fitted to the front wheel of all models with wire-spoked wheels. These components and assemblies are described in Chapter 5, but note that the UK XS250 SE and XS400 SE and the US XS400 SG, and SH models are equipped with a revised front brake caliper.

2 The new caliper differs mainly in the method of attachment to the caliper bracket, and this can be seen by comparing the accompanying line drawing to the earlier type shown in Fig. 5.2. The caliper removal process is slightly different, but this will be

self-evident during the operation. The caliper itself is a simple single-piston floating type, and the overhaul procedure is similar to that described in Chapter 5.

3   On later drum brake models the method of checking brake lining wear via a small inspection hole closed by a grommet was modified, and a wear indicator pointer was fitted to the brake arm. To assess wear, apply the brake and note the position of the pointer in relation to the scale on the brake plate. If it reaches the wear limit mark, dismantle the brake and renew the shoes as a pair.

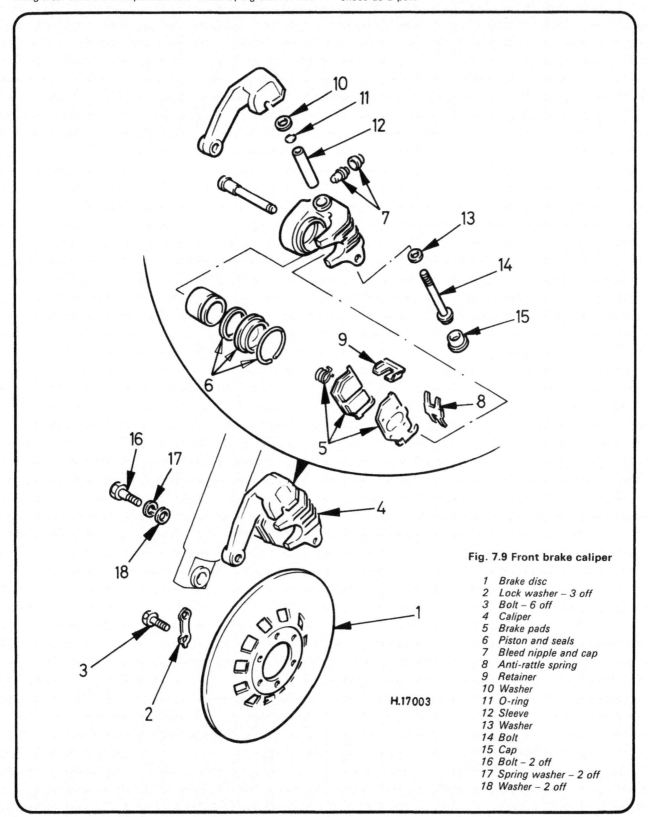

H.17003

Fig. 7.9 Front brake caliper

1   Brake disc
2   Lock washer – 3 off
3   Bolt – 6 off
4   Caliper
5   Brake pads
6   Piston and seals
7   Bleed nipple and cap
8   Anti-rattle spring
9   Retainer
10 Washer
11 O-ring
12 Sleeve
13 Washer
14 Bolt
15 Cap
16 Bolt – 2 off
17 Spring washer – 2 off
18 Washer – 2 off

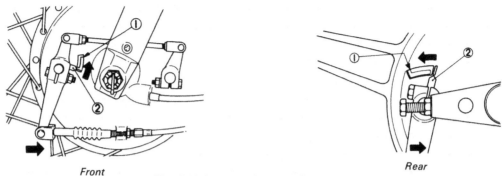

Front   *Rear*

Fig. 7.10 Drum brake wear indicator

1 *Wear limit*   2 *Pointer*

20 Electrical: system modifications – general

All later models received a number of detail modifications to the electrical system. Most of these are minor, such as revised bulb wattages or new headlamp or tail lamp designs to suit styling alterations. In the case of the former, such modifications are reflected in the specifications, whilst in the latter case, detail changes to mounting arrangements are self explanatory. Where fundamental changes have taken place, for example the change to an electronic rather than electromechanical voltage regulator, these are discussed in the sections which follow. Further information will be found in the wiring diagrams at the end of this Chapter.

21 Electronic voltage regulator: description and testing – UK models and US XS400 F (2L0), 2F (2V6), G (3F8), SG (3F9), H (4R5), SH (4R4), SJ (14V)

1 The above models were fitted with a sealed electronic voltage regulator in place of the earlier electromechanical type. The new unit is much smaller and lighter than the earlier version, and should normally be much more reliable in use. No adjustment is possible or required.
2 The operation of the regulator can be checked by connecting a dc voltmeter across the battery terminals. When the engine is run at more than 3000 rpm, a reading of 14.0 – 14.7 volts should be shown. If this is not the case, check that the battery is fully charged and in good condition and that the alternator and rectifier are working normally. If all are in good order, the voltage regulator can be assumed to have failed, and renewal will be required.
3 Note that care should be taken when dealing with any electronic component; an accidental short circuit can destroy such items in a few milliseconds. In particular, **never** run the engine with the battery disconnected. Always remove the battery when charging, and isolate the battery and alternator leads if carrying out any electric welding work. On no account apply high test voltages to the regulator terminals.

22 Starter interlock system: general – XS400 H, SH, SJ US models

The above models were equipped with a safety interlock arrangement designed to prevent the starter motor from operating unless the machine is in neutral or the clutch disengaged. The circuit, shown in the accompanying circuit diagram, has in addition to the normal starter components, a cut-off switch mounted on the clutch lever assembly, and a safety relay. It should be noted that a relatively insignificant fault, such as failure of the neutral switch or the interlock switch itself, can lead to the machine being impossible to start. In the event of the starter motor failing to operate, check these items in addition to the normal fault-finding procedures. If the fault persists it may be that the safety relay or its diode have failed internally, requiring renewal of the unit.

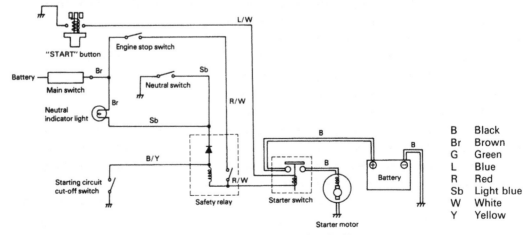

B	Black
Br	Brown
G	Green
L	Blue
R	Red
Sb	Light blue
W	White
Y	Yellow

Fig. 7.11 Starter interlock system circuit diagram

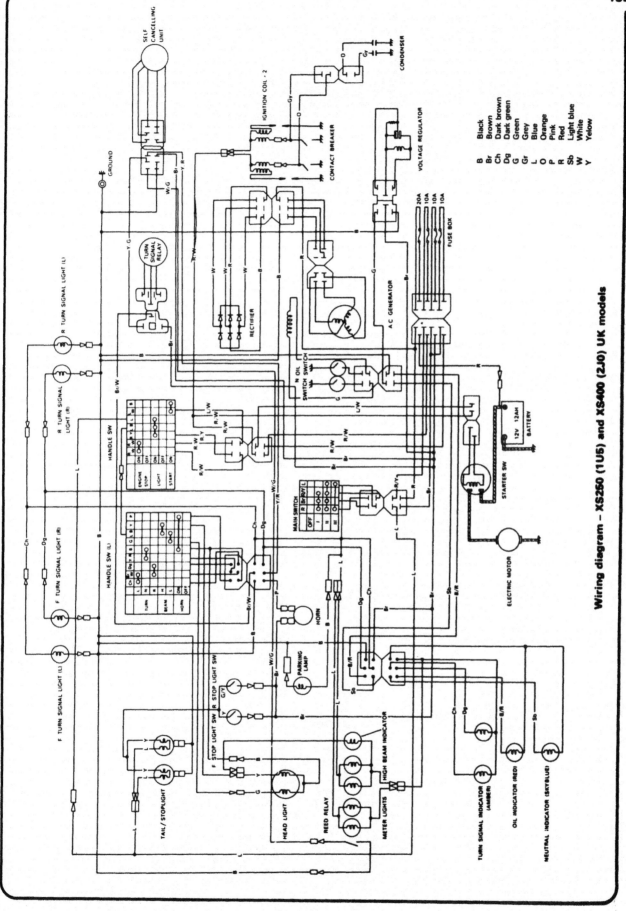

Wiring diagram – XS250 (1U5) and XS400 (2J0) UK models

B	Black
Br	Brown
Ch	Dark brown
Dg	Dark green
G	Green
Gr	Grey
L	Blue
O	Orange
P	Pink
R	Red
Sb	Light blue
W	White
Y	Yellow

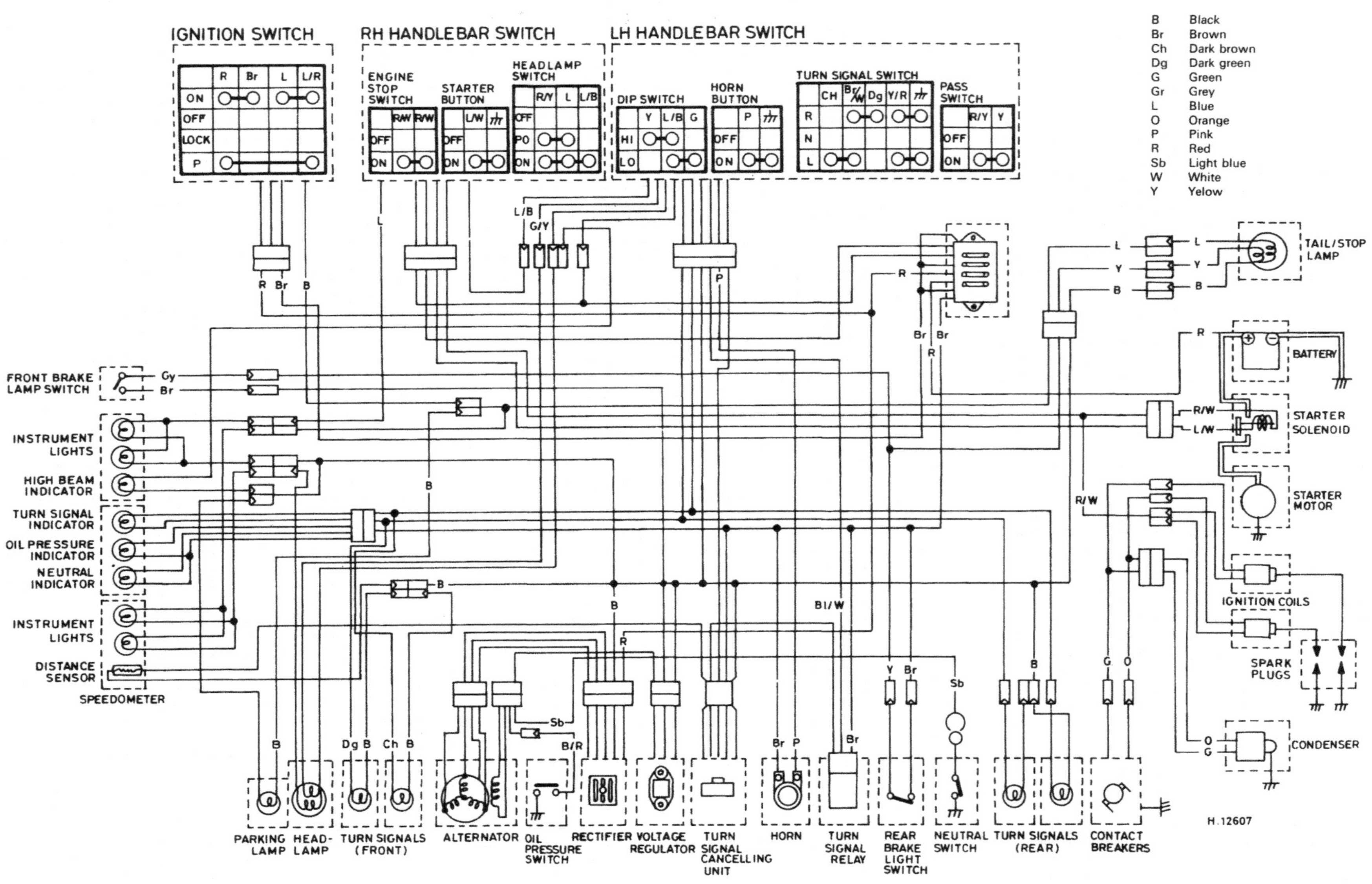

TAIL/STOP LAMP

BATTERY

STARTER SOLENOID

STARTER MOTOR

IGNITION COILS

SPARK PLUGS

CONDENSER

H.12607

B	Black
Br	Brown
Ch	Dark brown
Dg	Dark green
G	Green
Gr	Grey
L	Blue
O	Orange
P	Pink
R	Red
Sb	Light blue
W	White
Y	Yellow

CONTACT BREAKERS

TURN SIGNALS (REAR)

NEUTRAL SWITCH

REAR BRAKE LIGHT SWITCH

TURN SIGNAL RELAY

HORN

TURN SIGNAL CANCELLING UNIT

RECTIFIER VOLTAGE REGULATOR UNIT

OIL PRESSURE SWITCH

ALTERNATOR

TURN SIGNALS (FRONT)

HEADLAMP LAMP

PARKING LAMP LAMP

LH HANDLEBAR SWITCH

PASS SWITCH

TURN SIGNAL SWITCH

HORN BUTTON

DIP SWITCH

RH HANDLEBAR SWITCH

HEADLAMP SWITCH

STARTER BUTTON

ENGINE STOP SWITCH

IGNITION SWITCH

FRONT BRAKE LAMP SWITCH

INSTRUMENT LIGHTS

HIGH BEAM INDICATOR

TURN SIGNAL INDICATOR

OIL PRESSURE INDICATOR

NEUTRAL INDICATOR

INSTRUMENT LIGHTS

DISTANCE SENSOR

SPEEDOMETER

Wiring diagram – XS250 (3N6, 4A2) and XS400 (3N7) UK models

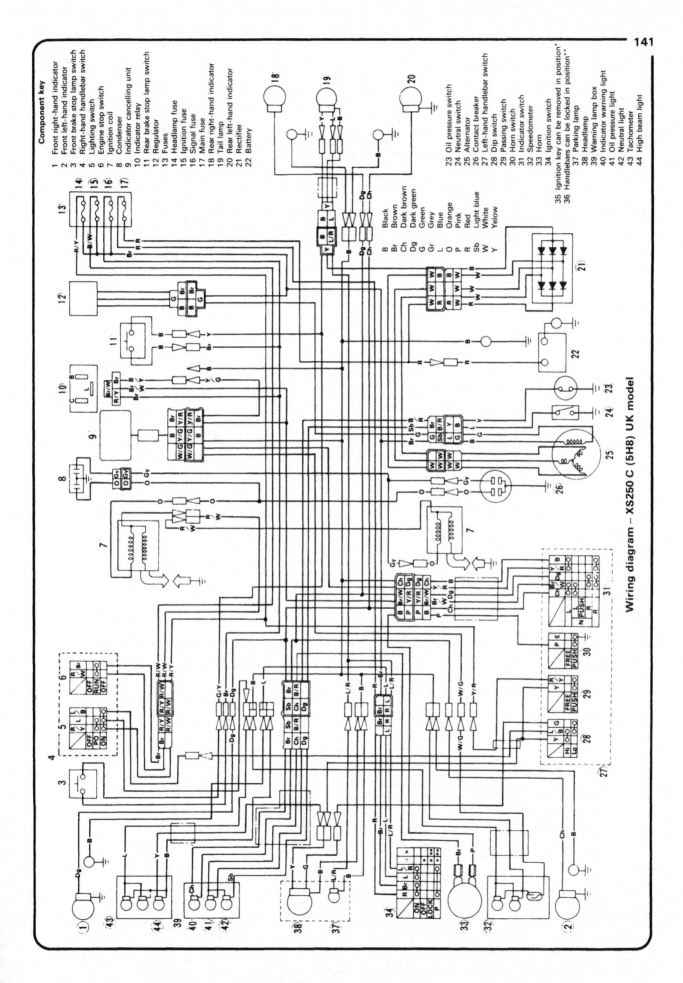

Component key

1 Front right-hand indicator
2 Front left-hand indicator
3 Front brake stop lamp switch
4 Right-hand handlebar switch
5 Lighting switch
6 Engine stop switch
7 Ignition coil
8 Condenser
9 Indicator cancelling unit
10 Indicator relay
11 Rear brake stop lamp switch
12 Regulator
13 Fuses
14 Headlamp fuse
15 Ignition fuse
16 Signal fuse
17 Main fuse
18 Rear right-hand indicator
19 Tail lamp
20 Rear left-hand indicator
21 Rectifier
22 Battery

23 Oil pressure switch
24 Neutral switch
25 Alternator
26 Contact breaker
27 Left-hand handlebar switch
28 Dip switch
29 Passing switch
30 Horn switch
31 Indicator switch
32 Speedometer
33 Horn
34 Ignition switch
35 Ignition key can be removed in position*
36 Handlebars can be locked in position**
37 Parking lamp
38 Headlamp
39 Warning lamp box
40 Indicator warning light
41 Oil pressure light
42 Neutral light
43 Tachometer
44 High beam light

B Black
Br Brown
Ch Dark brown
Dg Dark green
G Green
Gr Grey
L Blue
O Orange
P Pink
R Red
Sb Light blue
W White
Y Yellow

Wiring diagram – XS250 C (5H8) UK model

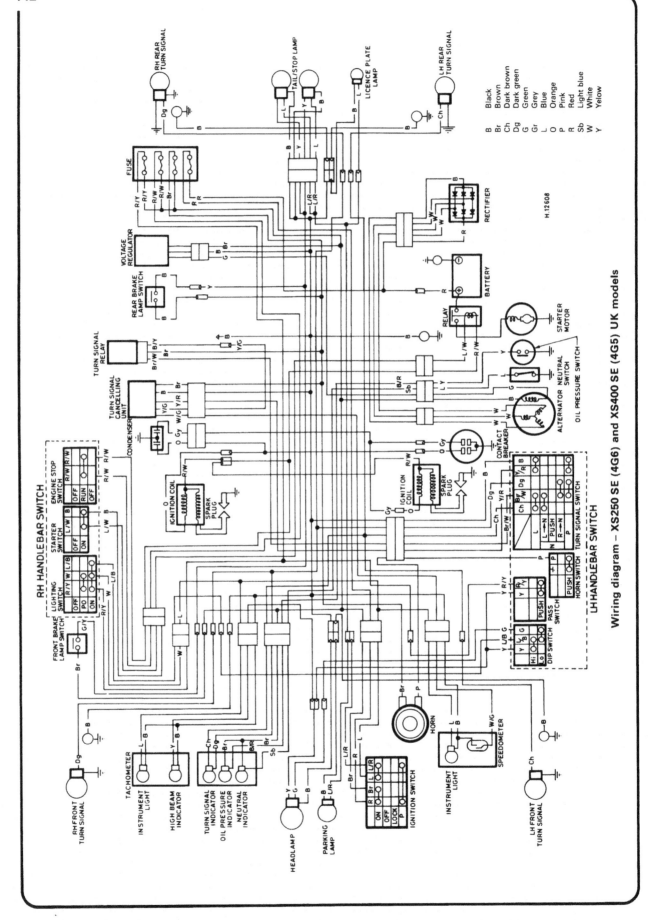

Wiring diagram – XS250 SE (4G6) and XS400 SE (4G5) UK models

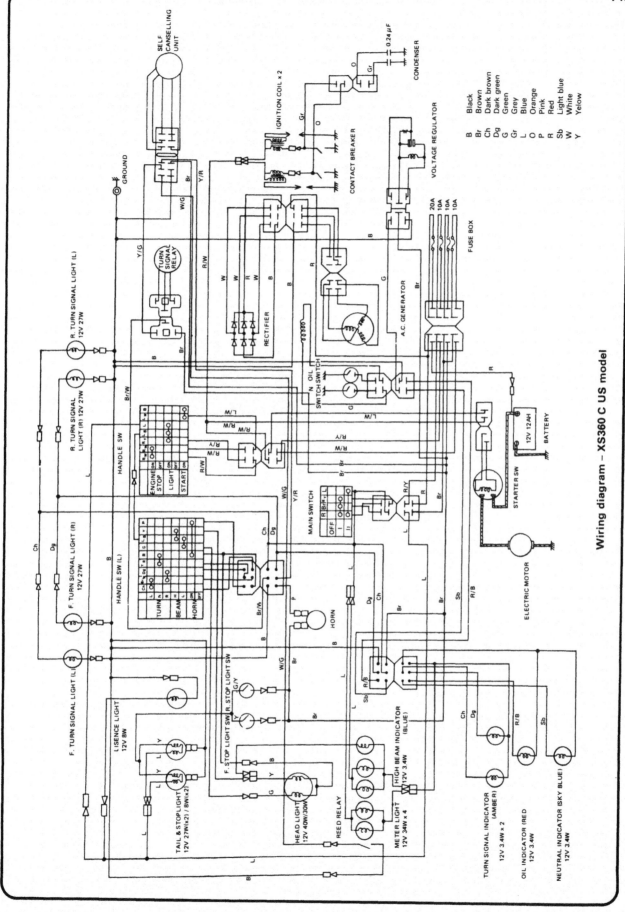

Wiring diagram – XS360 C US model

B	Black
Br	Brown
Ch	Dark brown
Dg	Dark green
G	Green
Gr	Grey
L	Blue
O	Orange
P	Pink
R	Red
Sb	Light blue
W	White
Y	Yellow

144

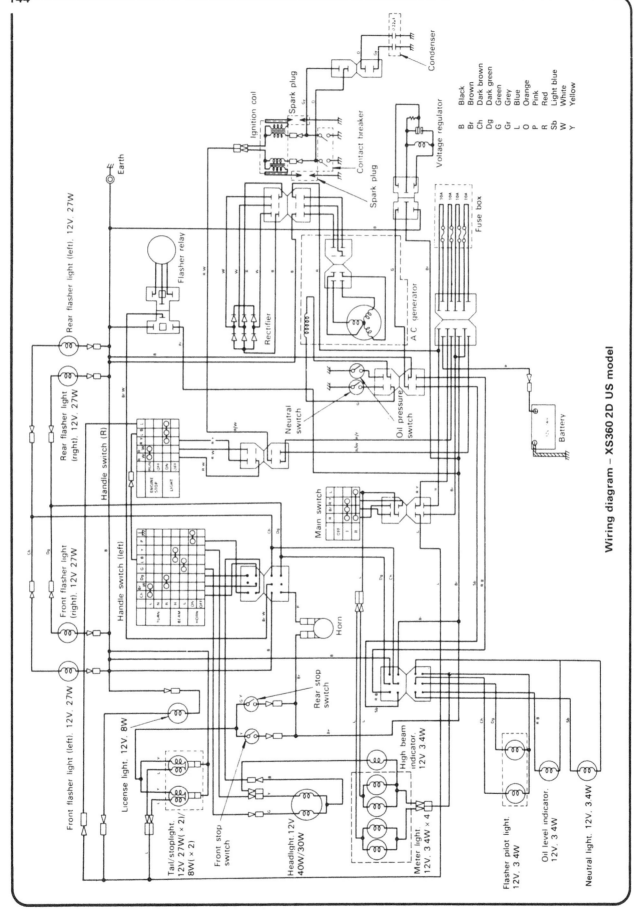

Wiring diagram – XS360 2D US model

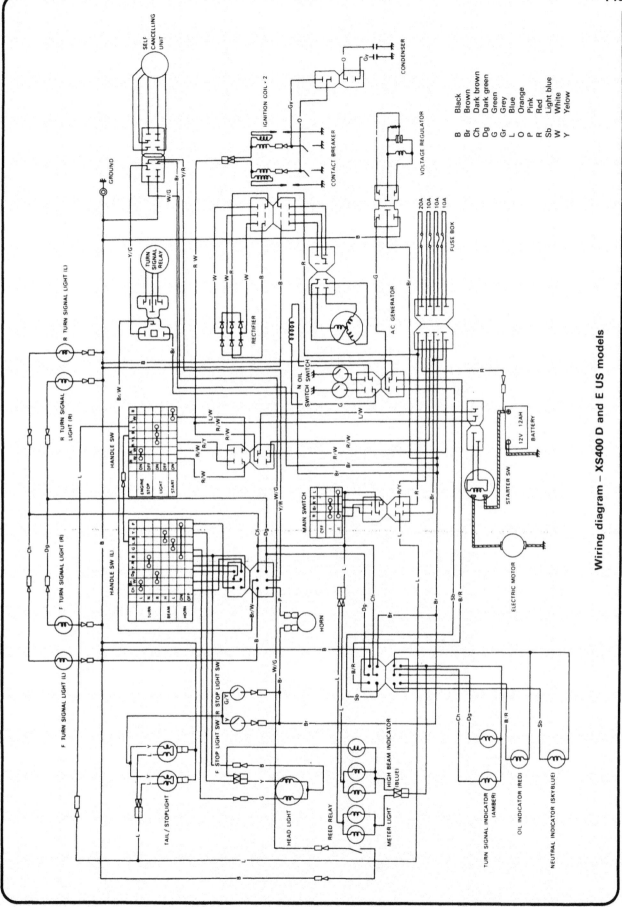

Wiring diagram – XS400 D and E US models

Wiring diagram – XS400 2E US model

B	Black
Br	Brown
Ch	Dark brown
Dg	Dark green
G	Green
Gr	Grey
L	Blue
O	Orange
P	Pink
R	Red
Sb	Light blue
W	White
Y	Yelow

* The key can be removed in this position

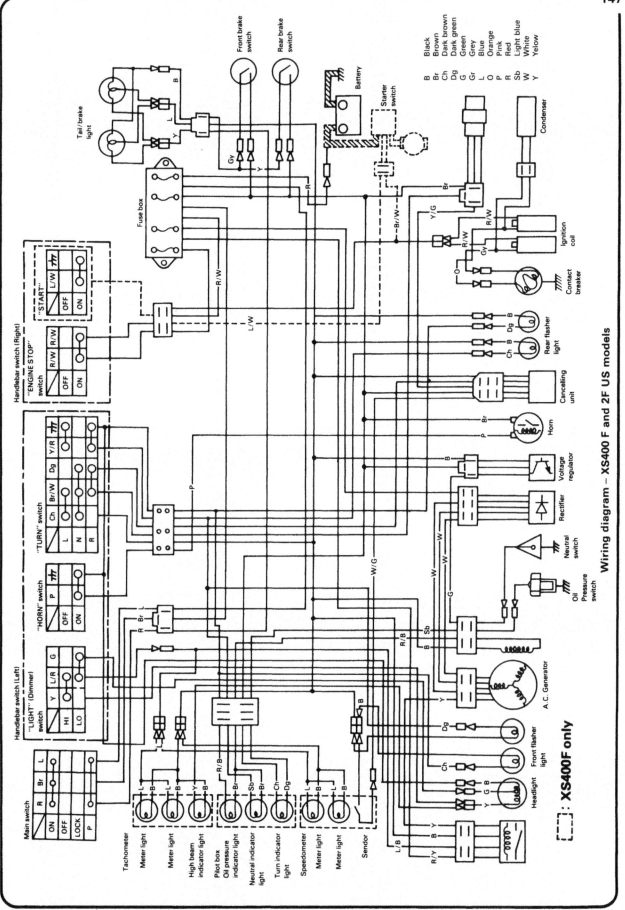

Wiring diagram – XS400 F and 2F US models

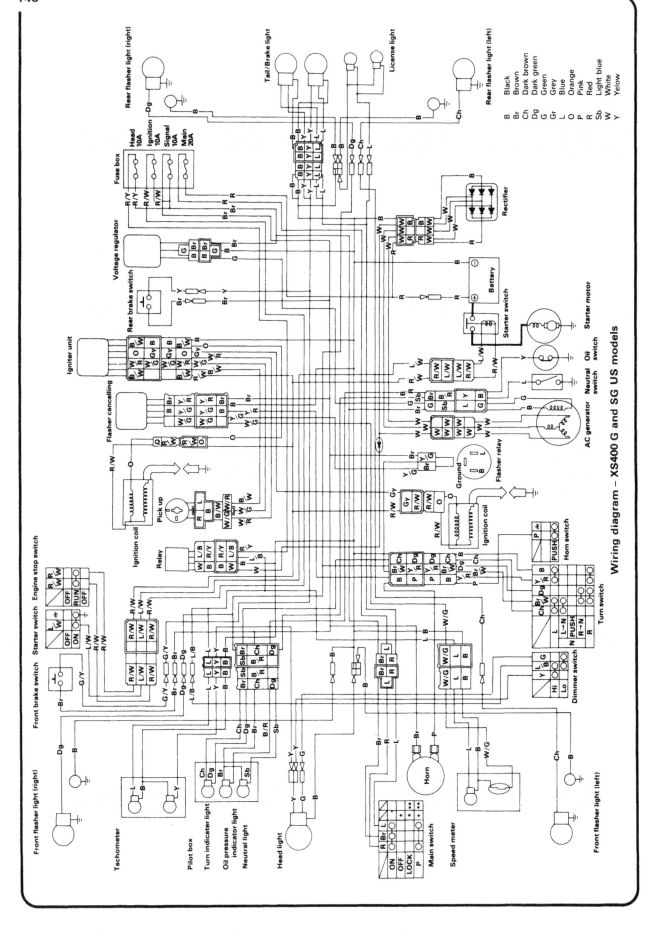

Wiring diagram – XS400 G and SG US models

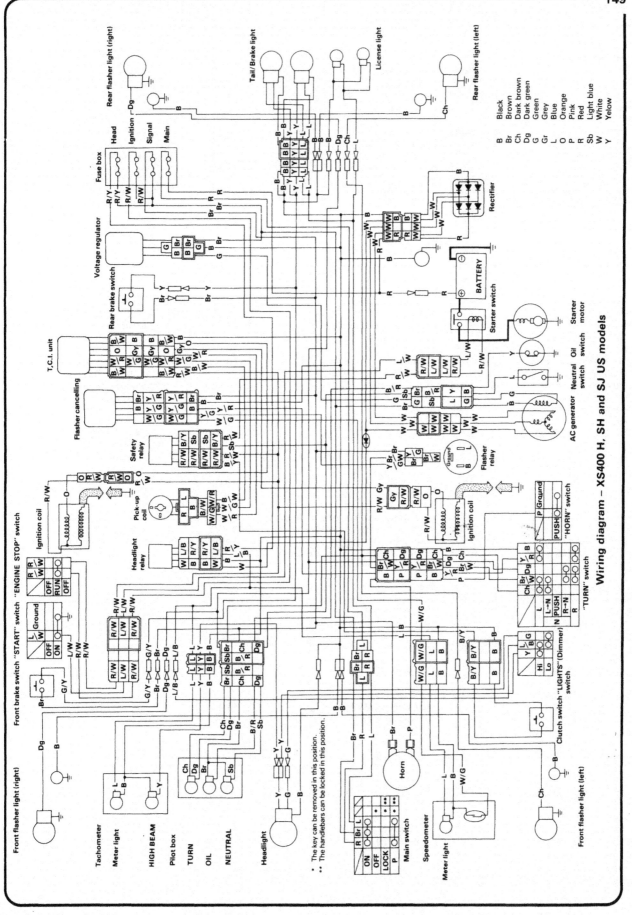

Wiring diagram – XS400 H, SH and SJ US models

Conversion factors

Length (distance)

Inches (in)	X	25.4	= Millimetres (mm)	X	0.0394	= Inches (in)
Feet (ft)	X	0.305	= Metres (m)	X	3.281	= Feet (ft)
Miles	X	1.609	= Kilometres (km)	X	0.621	= Miles

Volume (capacity)

Cubic inches (cu in; in³)	X	16.387	= Cubic centimetres (cc; cm³)	X	0.061	= Cubic inches (cu in; in³)
Imperial pints (Imp pt)	X	0.568	= Litres (l)	X	1.76	= Imperial pints (Imp pt)
Imperial quarts (Imp qt)	X	1.137	= Litres (l)	X	0.88	= Imperial quarts (Imp qt)
Imperial quarts (Imp qt)	X	1.201	= US quarts (US qt)	X	0.833	= Imperial quarts (Imp qt)
US quarts (US qt)	X	0.946	= Litres (l)	X	1.057	= US quarts (US qt)
Imperial gallons (Imp gal)	X	4.546	= Litres (l)	X	0.22	= Imperial gallons (Imp gal)
Imperial gallons (Imp gal)	X	1.201	= US gallons (US gal)	X	0.833	= Imperial gallons (Imp gal)
US gallons (US gal)	X	3.785	= Litres (l)	X	0.264	= US gallons (US gal)

Mass (weight)

Ounces (oz)	X	28.35	= Grams (g)	X	0.035	= Ounces (oz)
Pounds (lb)	X	0.454	= Kilograms (kg)	X	2.205	= Pounds (lb)

Force

Ounces-force (ozf; oz)	X	0.278	= Newtons (N)	X	3.6	= Ounces-force (ozf; oz)
Pounds-force (lbf; lb)	X	4.448	= Newtons (N)	X	0.225	= Pounds-force (lbf; lb)
Newtons (N)	X	0.1	= Kilograms-force (kgf; kg)	X	9.81	= Newtons (N)

Pressure

Pounds-force per square inch (psi; lbf/in²; lb/in²)	X	0.070	= Kilograms-force per square centimetre (kgf/cm²; kg/cm²)	X	14.223	= Pounds-force per square inch (psi; lbf/in²; lb/in²)
Pounds-force per square inch (psi; lbf/in²; lb/in²)	X	0.068	= Atmospheres (atm)	X	14.696	= Pounds-force per square inch (psi; lbf/in²; lb/in²)
Pounds-force per square inch (psi; lbf/in²; lb/in²)	X	0.069	= Bars	X	14.5	= Pounds-force per square inch (psi; lbf/in²; lb/in²)
Pounds-force per square inch (psi; lbf/in²; lb/in²)	X	6.895	= Kilopascals (kPa)	X	0.145	= Pounds-force per square inch (psi; lbf/in²; lb/in²)
Kilopascals (kPa)	X	0.01	= Kilograms-force per square centimetre (kgf/cm²; kg/cm²)	X	98.1	= Kilopascals (kPa)
Millibar (mbar)	X	100	= Pascals (Pa)	X	0.01	= Millibar (mbar)
Millibar (mbar)	X	0.0145	= Pounds-force per square inch (psi; lbf/in²; lb/in²)	X	68.947	= Millibar (mbar)
Millibar (mbar)	X	0.75	= Millimetres of mercury (mmHg)	X	1.333	= Millibar (mbar)
Millibar (mbar)	X	0.401	= Inches of water (inH₂O)	X	2.491	= Millibar (mbar)
Millimetres of mercury (mmHg)	X	0.535	= Inches of water (inH₂O)	X	1.868	= Millimetres of mercury (mmHg)
Inches of water (inH₂O)	X	0.036	= Pounds-force per square inch (psi; lbf/in²; lb/in²)	X	27.68	= Inches of water (inH₂O)

Torque (moment of force)

Pounds-force inches (lbf in; lb in)	X	1.152	= Kilograms-force centimetre (kgf cm; kg cm)	X	0.868	= Pounds-force inches (lbf in; lb in)
Pounds-force inches (lbf in; lb in)	X	0.113	= Newton metres (Nm)	X	8.85	= Pounds-force inches (lbf in; lb in)
Pounds-force inches (lbf in; lb in)	X	0.083	= Pounds-force feet (lbf ft; lb ft)	X	12	= Pounds-force inches (lbf in; lb in)
Pounds-force feet (lbf ft; lb ft)	X	0.138	= Kilograms-force metres (kgf m; kg m)	X	7.233	= Pounds-force feet (lbf ft; lb ft)
Pounds-force feet (lbf ft; lb ft)	X	1.356	= Newton metres (Nm)	X	0.738	= Pounds-force feet (lbf ft; lb ft)
Newton metres (Nm)	X	0.102	= Kilograms-force metres (kgf m; kg m)	X	9.804	= Newton metres (Nm)

Power

Horsepower (hp)	X	745.7	= Watts (W)	X	0.0013	= Horsepower (hp)

Velocity (speed)

Miles per hour (miles/hr; mph)	X	1.609	= Kilometres per hour (km/hr; kph)	X	0.621	= Miles per hour (miles/hr; mph)

Fuel consumption*

Miles per gallon, Imperial (mpg)	X	0.354	= Kilometres per litre (km/l)	X	2.825	= Miles per gallon, Imperial (mpg)
Miles per gallon, US (mpg)	X	0.425	= Kilometres per litre (km/l)	X	2.352	= Miles per gallon, US (mpg)

Temperature

Degrees Fahrenheit = (°C x 1.8) + 32

Degrees Celsius (Degrees Centigrade; °C) = (°F - 32) x 0.56

It is common practice to convert from miles per gallon (mpg) to litres/100 kilometres (l/100km), where mpg (Imperial) x l/100 km = 282 and mpg (US) x l/100 km = 235

English/American terminology

Because this book has been written in England, British English component names, phrases and spellings have been used throughout. American English usage is quite often different and whereas normally no confusion should occur, a list of equivalent terminology is given below.

English	American	English	American
Air filter	Air cleaner	Number plate	License plate
Alignment (headlamp)	Aim	Output or layshaft	Countershaft
Allen screw/key	Socket screw/wrench	Panniers	Side cases
Anticlockwise	Counterclockwise	Paraffin	Kerosene
Bottom/top gear	Low/high gear	Petrol	Gasoline
Bottom/top yoke	Bottom/top triple clamp	Petrol/fuel tank	Gas tank
Bush	Bushing	Pinking	Pinging
Carburettor	Carburetor	Rear suspension unit	Rear shock absorber
Catch	Latch	Rocker cover	Valve cover
Circlip	Snap ring	Selector	Shifter
Clutch drum	Clutch housing	Self-locking pliers	Vise-grips
Dip switch	Dimmer switch	Side or parking lamp	Parking or auxiliary light
Disulphide	Disulfide	Side or prop stand	Kick stand
Dynamo	DC generator	Silencer	Muffler
Earth	Ground	Spanner	Wrench
End float	End play	Split pin	Cotter pin
Engineer's blue	Machinist's dye	Stanchion	Tube
Exhaust pipe	Header	Sulphuric	Sulfuric
Fault diagnosis	Trouble shooting	Sump	Oil pan
Float chamber	Float bowl	Swinging arm	Swingarm
Footrest	Footpeg	Tab washer	Lock washer
Fuel/petrol tap	Petcock	Top box	Trunk
Gaiter	Boot	Torch	Flashlight
Gearbox	Transmission	Two/four stroke	Two/four cycle
Gearchange	Shift	Tyre	Tire
Gudgeon pin	Wrist/piston pin	Valve collar	Valve retainer
Indicator	Turn signal	Valve collets	Valve cotters
Inlet	Intake	Vice	Vise
Input shaft or mainshaft	Mainshaft	Wheel spindle	Axle
Kickstart	Kickstarter	White spirit	Stoddard solvent
Lower leg	Slider	Windscreen	Windshield
Mudguard	Fender		

Index

Preserving Our Motoring Heritage

< The Model J Duesenberg Derham Tourster. Only eight of these magnificent cars were ever built – this is the only example to be found outside the United States of America

Almost every car you've ever loved, loathed or desired is gathered under one roof at the Haynes Motor Museum. Over 300 immaculately presented cars and motorbikes represent every aspect of our motoring heritage, from elegant reminders of bygone days, such as the superb Model J Duesenberg to curiosities like the bug-eyed BMW Isetta. There are also many old friends and flames. Perhaps you remember the 1959 Ford Popular that you did your courting in? The magnificent 'Red Collection' is a spectacle of classic sports cars including AC, Alfa Romeo, Austin Healey, Ferrari, Lamborghini, Maserati, MG, Riley, Porsche and Triumph.

A Perfect Day Out

Each and every vehicle at the Haynes Motor Museum has played its part in the history and culture of Motoring. Today, they make a wonderful spectacle and a great day out for all the family. Bring the kids, bring Mum and Dad, but above all bring your camera to capture those golden memories for ever. You will also find an impressive array of motoring memorabilia, a comfortable 70 seat video cinema and one of the most extensive transport book shops in Britain. The Pit Stop Cafe serves everything from a cup of tea to wholesome, home-made meals or, if you prefer, you can enjoy the large picnic area nestled in the beautiful rural surroundings of Somerset.

> John Haynes O.B.E., Founder and Chairman of the museum at the wheel of a Haynes Light 12.

< The 1936 490cc sohc-engined International Norton – well known for its racing success

The Museum is situated on the A359 Yeovil to Frome road at Sparkford, just off the A303 in Somerset. It is about 40 miles south of Bristol, and 25 minutes drive from the M5 intersection at Taunton.
Open 9.30am - 5.30pm (10.00am - 4.00pm Winter) 7 days a week, *except Christmas Day, Boxing Day and New Years Day*
Special rates available for schools, coach parties and outings Charitable Trust No. 292048